Science in a Social Context

Are Science and Technology Neutral?

Joan Lipscombe
Siscon Research Fellow,
University of Leeds

and

Bill Williams
Director of Combined Studies in Science,
University of Leeds

Butterworths
LONDON - BOSTON
Sydney - Wellington - Durban - Toronto

The Butterworth Group

United Kingdom	Butterworth & Co (Publishers) Ltd
London	88 Kingsway, WC2B 6AB
Australia	Butterworths Pty Ltd
Sydney	586 Pacific Highway, Chatswood, NSW 2067 Also at Melbourne, Brisbane, Adelaide and Perth
Canada	Butterworth & Co (Canada) Ltd
Toronto	2265 Midland Avenue, Scarborough, Ontario, M1P 4S1
New Zealand	Butterworths of New Zealand Ltd
Wellington	T & W Young Building, 77–85 Customhouse Quay, 1, CPO Box 472
South Africa	Butterworth & Co (South Africa) (Pty) Ltd
Durban	152–154 Gale Street
USA	Butterworth (Publishers) Inc
Boston	19 Cummings Park, Woburn, Mass. 01801

First published 1979
ISBN 0 408 71312 7

British Library Cataloguing in Publication Data

Lipscombe, Joan
Are science and technology neutral?
1. Science – Social aspects 2. Technology – Social aspects
I. Title II. Science in a Social Context (Project)
301.24'3 Q175.5 78-41267

ISBN 0-408-71312-7

Typeset by Butterworth Litho Preparation Department
Printed in England by Billing and Sons Ltd,
Guildford and London

Introduction

The controversy which surrounds the questions of whether Science and Technology can be considered neutral has strong supporters on both sides. In this book we introduce you to some of the arguments used by people to support a particular point of view and consider some of the questions which are raised as soon as we begin to think carefully about what is implied by the idea that science and technology are neutral. We have made no attempt to define 'science', 'technology' or 'neutrality'. The reason for this is that all three words mean different things to different people and many of the arguments used are related to one person's particular interpretation of the meaning of these words. You may feel it is impossible to consider this question without giving specific definitions. One consequence of doing this, however, would be to narrow the field of enquiry considerably. We would have to investigate how a closely defined concept measures up to specific criteria, themselves equally tightly defined. It might then be possible to come to a precise conclusion: to 'prove' whether or not science and technology are neutral. In reality, though, words such as 'science' and 'technology' or 'scientist', 'technologist' and 'engineer' are not used with clear and limited meaning. They are frequently seen as interchangeable and each word may be used successively to describe the same activity or person. Similarly, the boundaries which distinguish the body of knowledge called 'science' from other bodies of knowledge are not at all clear. Definitions which rely on the methods used have other limitations. There is a danger that, whatever definition is chosen, most of what is generally considered to come under the umbrella of 'science and technology' will be excluded from the conclusion reached. Our taking such a definition would not prevent others from using the terms more broadly and transferring conclusions about a tightly defined 'science' or 'technology' to all activities commonly called 'science and technology'. We have taken a wide range of interpretations and arguments to help you formulate your own opinion and examine your own beliefs about the issues involved.

Why should the ordinary scientist be concerned about such a debate? Surely it is of limited interest, being of importance only to philosophers? We think not. In a highly complex technological society where science and its related developments impinge on every aspect of life it is important that every scientist should be quite clear in his own mind what his conception of the role of science in society is and whether this is realistic in terms of what actually happens. The common, often unspoken, belief among many scientists and technologists is that science is neutral – neutral in a very complete sense of the word: not ideologically based, an absolute, the truth, objective, there for all to determine the facts, theories arising naturally out of the facts, the same facts and theories accessible to all men and acceptable to all men at all times. The very act of stating this assumption, making it explicit, sows the seed of doubt in the minds both of the writer and the reader: can any human activity be so perfect as this suggests? And the immediate answer which springs to mind: 'of course not, we well know that observations are constantly being improved and added to, old theories are being refined and replaced by new theories . . . and so on' is almost equally immediately seen as not good enough. Not good enough because this answer assumes that, underlying man's inadequate view of nature, there is a 'perfect' view which we are approaching step by step which *will* have the complete, perfect, neutral character which our current state of knowledge presently lacks; that the detectable imperfections in science's completeness are of a temporary nature and will, in the course of time, gradually be ironed out. Is this asymptotic (approaching but never reaching) model of the perfection, the completeness, of science any more tenable than the temporary, provisional, model which it was erected to defend?

If this picture of the neutrality of science is commonly, unthinkingly, held by most scientists and technologists, how much more so is an even more idealized view held by the layman, the non-scientist, who, overawed by the evidence of the success of science and its applications which he sees all around him, accepts this neutral view of science with the omniscience which it suggests, and elevates the scientist to the role of high priest expounding its truths. It is arguable that the certainty which this view of science provides for the 'ordinary man' is necessary to replace the certainty which religion hitherto provided and which the advances of science have badly dented if not shattered. Whether that is true or not, this attitude to science reinforces the scientist's belief in scientific neutrality, the self-contained completeness of science. It encourages the scientist to bask in the reverence afforded to him as he propounds these great, immutable truths to the lay public, a reverence which he would certainly not be accorded if science were seen to be more changing, more fallible, more socially based.

Thus, in this way, the non-scientist forms a view which is a stereotype (a generalized image he applies to cover a particular category of things or people). On the basis of this image the general public gives tacit agreement to immense funds being poured into scientific research and development, in part because an aura of objectivity and truth surrounds pronunciations of scientists. Is this a realistic view of science, and, if not, should scientists attempt to change it?

Chapter One is concerned mainly with considering the neutrality of science in the context of what is normally defined as pure and applied science. We look at the way people's concepts of science have varied over the years, as well as the current thinking on this topic. We examine three main aspects – science as a body of knowledge; science's aims and its methodology; science as an activity in terms of what scientists actually do – and look at some of the arguments that people have used to support both sides of the neutrality controversy. The last section of this chapter raises the question of how the issue of neutrality affects a further issue – that of whether scientists have a social responsibility. Our approach is to try to show that arguments relating to social responsibility often depend on a prior conclusion about whether or not science is neutral. The question of social responsibility in science is also considered in the series of Siscon Units *The Scientist's Social Responsibility* by Judith Hargreaves, to whom we are indebted for advice on this matter.

In Chapter Two we turn our attention to technology. The border between science and technology is no longer clear. Technology is certainly related to science even though the exact relationship is not known precisely. Many people trained as scientists are working directly in technology and the applications of scientific research affect people and society directly. How far can technology, as the application of science, be considered neutral?

Chapter Three leads on to a specific situation. We use the current trend promoting intermediate technology in the Third World as an example to illustrate some of the issues which have been raised in the previous chapters.

Note on readings

Readings for this book are divided into two main types, those which are central to the text of the book and can be regarded as essential and those which are useful as additional readings either as background information to the discussion/essay topics suggested or in broadening the students' understanding and knowledge of the issues involved. The essential readings are listed at the beginning of the relevant chapter. Some of the additional readings are listed with the particular discussion/essay topic for which they are most helpful whilst the remaining ones are listed at the end of each chapter under the title of further reading. These suggestions are in no way exclusive and many similar ones discussing the nature of science and technology and the role of scientists and technologists can be found in books about the philosophy or sociology of science and in general science and social science journals. Of particular interest in the context of this book would be articles by practicing scientists which illustrate their views on the issues we have raised.

Chapter One
Science

ESSENTIAL READING

1.1 Chain, Sir Ernest (1970). 'Social Responsibility and the Scientist'. *New Scientist*, 22 October, pp. 166–70.
1.2 Black, M. 'Is Scientific Neutrality a Myth?', a lecture delivered to the Annual General Meeting of the American Association for the Advancement of Science, 27 January 1975 (reprinted as Appendix One).
1.3 Rose, S. and Rose, H. (1971). 'The Myth of the Neutrality of Science'. In *The Social Impact of Modern Biology.* Edited by Watson Fuller. London, Routledge & Kegan Paul, and In *Impact of Science on Society* **21**, No. 2, 1971.

A. Scientific knowledge

Sir Ernest Chain states clearly the thesis of scientific neutrality in the context of science as a body of knowledge:

> . . . science, as long as it limits itself to the descriptive study of the Laws of Nature, has no moral or ethical quality, and this applies to the physical as well as the biological sciences.[1]

It is often said that science seeks to ascertain the truth about nature. Its hypotheses aim to move nearer and nearer to an accurate description of natural laws, which are seen as universal truths. Objective reasoning cannot deny scientific facts and all scientists must inevitably reach the same conclusion. There is room for controversy and dissent, but only on the basis of facts which do not appear to fit into a particular hypothesis. Value judgements, cultural biases or political standpoints do not in any way influence or determine scientific knowledge. There is nothing 'good' or 'bad' about scientific knowledge. Galileo sets this view out quite clearly:

> If this point of which we dispute were some point of law, or other part of the studies called the humanities wherein there is neither truth nor falsehood, we might give sufficient credit to the acuteness of wit, readiness of answers, and the greater accomplishment of writers, and hope that he who is most proficient in these will make his reason more probable and plausible. *But the conclusions of natural science are true and necessary, and the judgement of man has nothing to do with them.*[2]

This view has been carried forward and is strongly supported today (see for example Bronowski[3] who attributes to science "an unrelenting independence in the search for truth that pays no attention to received opinion or expediency or political advantage").

This idea of the neutrality of scientific knowledge is now being challenged in several ways. The Roses consider that it is challenged by the very methods which scientists use. Not only does science policy effectively dictate the problems to which science's resources are directed but the paradigms (the generally accepted fundamental beliefs about a particular phenomenon which describe its nature, explain experimental results and define the further areas of investigation which can proceed without challenging the basic assumptions) which are used tend to limit the range of answers which are possible. They imply the existence of certain relationships and research is directed to proving these relationships. The idea that such relationships might exist and thoughts about what those relationships could be are influenced by social and political factors. The Roses also illustrate that the metaphors (the use of a descriptive term in a situation where it is not *literally* applicable) used to express scientific concepts are value laden and reflect the value systems of the society at that time. Both these latter points are well illustrated in one of the examples which the Roses use:

> Anyone reared on the biochemistry that developed from the 1930s will recall that the central theme of teaching and research was that of energy. The key to workings of the cell – referred to as the cellular economy – was the flow of energy within the cell and the clue to this energy flow was provided by a substance known as ATP. It was Lipman in the early 1940s who produced the key metaphor for the activity of ATP, it was the energy *currency* of the cell. Oxidation of glucose resulted in the synthesis of ATP – storing it in an 'energy bank'. A compound related to ATP, creatine phosphate or CP, is synthesized when there is an abundance of ATP; it is referred to as an energy 'deposit-account' compound compared with the ATP 'current-account'. Energy, economy, banks, current and deposit accounts . . . such terms are more than mere puns; they both reflect the vision that the cell biologist has of his phenomena and help to direct his thoughts to new experiments. Much of the biochemistry of the 1930s and 1940s through to the mid-1950s was conducted within this type of language system, with its pre-Keynesian undertones.
>
> From the mid-1950s on, though, the language system began to change. The central metaphor was altered; the paradigm switched . . . What are today's central metaphors? . . . They are those of control, community and communication, feedback, interaction, repression and regulation, switching on and switching off.
>
> These are the metaphors of cybernetics and the engineering

and computing sciences; but still more are they metaphors of today's managerial, post-Keynesian capitalism or bureaucratic socialism. They form the scaffolding for today's puzzle-solving research in biology. And in their turn they serve to bolster the structure of society by providing it – by virtue of a measure of determined extrapolation – with a biological rationale as well. The role of such metaphors cannot be simply subsumed as punning; they frame too closely the thought processes of the researcher.[4]

A similar argument is used by Young.[5] He looks at the evolutionary debate during the nineteenth century and examines how theological, ethical, ideological and other 'non-scientific' arguments were closely intertwined with scientific and pseudo-scientific (falsely appearing to be scientific) ones. It was impossible to separate them and stand-points on 'the reality' were determined by considering all these aspects. What people were prepared to accept as the 'truth' was not determined by science alone. In this way the thesis of the neutrality of scientific knowledge is challenged by the subtle and often unacknowledged influence of social factors.

Another challenge is associated with the deliberate suppression of scientific knowledge or the active promotion of particular theories which conform with a specific political situation. The Lysenko[6] affair in Russia is often quoted as an extreme example of this. A whole area of genetics was eliminated from Russian teaching and Lysenko's theories imposed because they were more supportive to the political system. Russian scientists worked within the framework of his theories, believing them to be 'true', at least as far as the existing evidence was concerned. Other examples can be found to illustrate the search for scientific theories which bolster up a particular political or social system.[7] The point here is that in theory science may be concerned with the search for the ultimate truth, but in reality the body of accepted scientific knowledge at any particular moment is closely related to the existing social system and, in the extreme case, 'scientific facts' may be the invention of a political regime.

Another aspect of this controversy in the context of scientific knowledge is the alleged impossibility of scientific knowledge giving rise to normative or evaluative statements. Science, say the proponents of this neutrality thesis, concerns itself purely with a description of the world as it is. It cannot give rise to statements about what should or should not be (normative), nor can it pass judgement on what is good or bad (evaluative). It is an orthodox philosophical argument that the only valid conclusions of deductive arguments are the ones which contain only material which is already in the premises. Consequently scientific premises (factual) cannot lead to normative or evaluative statements. For example it is possible for science to provide factual statements about the effects of dropping a nuclear bomb. It can predict what area will be affected, what proportion of the population will be killed outright, what proportion will die as a result of injuries and radiation and so on. It cannot say, however, whether it is right or wrong to drop a nuclear bomb. Such a view relies on a value judgement about the relative importance of the political benefits which might follow from such an action and the destruction and human misery which it would cause. Science has nothing to say about this, if science is neutral. This viewpoint is attacked by Black.[8] He argues that it falls down because of the problem of identifying which premises are factual. A normative statement (he uses the example of "murder is a sin") can be expressed in the same way as a factual one and there are considerable difficulties in clearly distinguishing one from the other. Furthermore he asserts that some normative evaluative propositions are objective (generally accepted and not subject to individual values) and this removes the distinction which separates scientific propositions from others. The inclusion of generally acceptable normative statements would allow normative conclusions to follow from scientific premises. An example of this could be:

> Plant defoliants can cause food shortage (factual).
> Food shortages lead to people starving (factual).
> It is wrong that people should starve directly because of man's action (normative).
> Therefore plant defoliants should not be used (normative).

Discussion/essay suggestions

A.1 Some of the opposition to the concept of neutrality of science is based on examination of how scientists behave, e.g. Lysenko, or the reaction to Darwin's theories. How important is this criticism? Should the 'weakness' of human beings in allowing other considerations to influence them to be allowed to destroy the ideal of scientific neutrality?

See:

1.4 Young, R. M. (1971). 'Evolutionary Biology and Ideology – Then and Now'. In *The Social Impact of Modern Biology.* Edited by Watson Fuller. London, Routledge & Kegan Paul.

1.5 Caspari, E. W. and Marshak, R. E. (1965). 'The Rise and Fall of Lysenko'. *Science,* 16 July, pp. 275–278.

A.2 M. Black indicated that he believes that there are some objectively true normative statements (*see* Appendix One, page 40). Can you think of any such statements and how far can these be regarded as 'generally accepted and not subject to individual values'?

A.3 Make a list of some of the metaphors which are used in your own discipline (remember the examples cellular economy and energy currency given by the Roses). What reasons can you give for using each of your examples?

A.4 Do you think that scientists sometimes hope that one theory rather than another will turn out to be correct? If you do, give examples to support your view. Could such hopes influence the way in which the scientist interprets data? How could this happen in practice?

B. Scientific aims and methods in the context of neutrality

Neutrality associated with scientific aims implies that neither good nor evil is intended from the results of scientific investigation; the aim is the pursuit of knowledge in itself. The argument is often used to defend the so-called rights of scientists to research into what interests them with no outside interference or control. There is, however, an inbuilt implication within this argument, that the pursuit of knowledge is in itself a good thing. Black[8] draws a distinction between the pursuit of knowledge as information and knowledge as understanding. He points out that the collection of information in itself is a product of value judgements. Choices about what information to collect, what information to ignore, are value laden. Similarly he claims that it may be better to remain ignorant than to understand some things. Do we really want to know the nutritional value of human flesh or understand the processes necessary to convert human excreta into edible food? It is often not possible to isolate, in the real world, the understanding from the purpose which that understanding is to serve.

What does history tell us of the neutrality of science in this sense? Examination of the past reveals that it has long been recognized that science (or at least its application) could be a power for good or evil. This was recognized by Francis Bacon (1561–1626) who not only saw science as a means to the end of improving man's lot, but who also advocated the suppression of some scientific knowledge to prevent its misuse by the government. Science has often been seen as the means of relieving human burdens and this, and not the disinterested pursuit of knowledge, has often motivated scientists. The values of scientists in deciding priorities for their research and in deciding on what problems science should be brought to bear affect the issue of the neutrality of science.

How do the methods of science fit into the neutrality thesis? Scientific method is fairly well defined. It involves conducting investigations under carefully controlled conditions and much of the validity of scientific knowledge rests on the reproduceability of the results. Methodology may be neutral in the sense that the findings of such methods are unrelated to the value system of the researcher, although this is challenged by some schools of thought. (Students who wish to pursue this aspect should see Siscon Unit, *Science and Rationality*.) It can, however, be argued that both the choice of methods and the choice of problem, what is permissible in the name of scientific research, is value laden and influenced by the ethical and moral state of the society in which the research is being conducted. We can use vivisection to illustrate this point. This has been accepted as an important research technique in the quest for medical knowledge. Some scientists working in this field have accepted the use of vivisection only because they see the goal of eliminating disease as justifying the abhorrent methods which they have to use. It is quite clear that value systems are operating here, both in the choice of goal and the methods to be used. Charles Darwin regarded the disinterested pursuit of knowledge as an insufficient end to justify such means and such research methods could only be defended because of the over-riding importance of eliminating disease. The practice of vivisection was to him totally unacceptable if it was solely to satisfy "mere damnable and detestable curiosity". Use of experimentation on animals has today been extended far beyond the bounds of pure medical research[9].

Darwin clearly had a view that species other than *homo sapiens* had a 'right' to their place in the universe and that the assumption that the interests of members of other species could justifiably be ignored to satisfy the curiosity of the human race was – at best – 'not proven'. But, at this point, it is worth pausing to extend this enquiry: *is* medical research an adequate excuse for vivisection? The basis on which such justification is generally – usually unquestioningly – accepted is the hierarchical view of nature, a view which, in one form or another, goes back as far as Plato. In its modern form man is at the top of the scale of life and other species occupy lower and lower points on the scale with correspondingly less and less rights and – in particular – less and less immunity from experimentation. So, for example, the very rare experiments on man, whether in a German concentration camp or in an American prison, with or without the agreement of the subjects, always gives rise to an outcry; the higher apes other than man are seldom subjected to surgery, pickling in tobacco smoke, drugging or other less refined forms of torture and, when they are, are treated according to a set of rules which accord them a certain 'respect'.

Octopuses, cockroaches and frogs cause us less concern – genetic engineering with possible ill effects for humans has suddenly become a matter of great concern for the scientific community and has provided the occasion for a brave display of scientific responsibility, but genetic engineering has been practised on frogs for a long time without anyone worrying very much – the lower down the scale, the less it matters. Dogs form an interesting exception to the rule. Dogs are often used as experimental animals (interesting euphemism), are carved up, drugged, smoked, dismembered, roughly in accord with their position in the hierarchy, but despite the apparent 'equity' are singled out by the anti-vivisectionists as a focus for their indignation. The recent furore, caused by the discovery that the tobacco researchers were using beagles in their attempts to make cigarette smoking safe for human beings springs to mind. Why? Why protest about dogs and not rabbits or guinea pigs or rats? Is it perhaps that the dog's close association with man somehow entitles it to a leg up the scale?

We have come to accept the anthropocentric (man centered) view of the universe more or less without question. But it is interesting and revealing to adopt another viewpoint and see where this leads us. So for example Faulkner[10] suggests, amusingly, that the cat is the lord of the universe, skilfully using men (and women) to provide food, shelter and all the creature comforts that it requires. What if we were to extend this idea further? What would the world look like viewed by say, the grey seal, fighting hard for survival when stocks of fish are depleted by man's over-fishing, in seas which are polluted by man's wastes – both arguably the consequences of man's over-population – and where, despite these and other handicaps – the seal population is culled (mature animals and pups are battered to death with clubs or skinned alive) in order to keep their number 'under control'. In a seal-dominated world who would be culling whom? Or how would the world look to the creature – say a bird of prey – at the end of a food chain in which DDT is progressively concentrated? Food poisoned by man's activities, survival threatened by DDT's effects on fertility, and hunted and persecuted by farmers who believe – often without justification – that by fulfilling its basic need for food in the only way it knows how, it steals their 'possessions'.

Clearly then a different value system would lead to a different view of what scientific methods and aims are acceptable. People who believe that all life is sacred and would allow nothing to be done to harm or injure any form of life are regarded as being odd, but if their view predominated a different pattern of scientific research methods would emerge.

Discussion/essay suggestions

B.1 It is often said that scientists should be free to follow their research interests without outside interference. They alone should determine what are their priorities and money should be made available to finance their research. In what circumstances would this be a reasonable claim and how far do the realities of today's world meet these circumstances?

See:

1.6 Bronowski, J. (1971). 'The Disestablishment of Science'. In *The Social Impact of Modern Biology*. Edited by Watson Fuller. London, Routledge & Kegan Paul.

B.2 Darwin provides us with one example, vivisection, of how, in his view, the pursuit of knowledge in itself is insufficient to justify a scientific technique. What other examples can you think of where a subjective judgement about the importance of the practical results of scientific research is used to overcome ethical objections to the means to be used? Psychological and medical research are obvious fields to look for such examples, but you may find some others in the physical sciences. If these ethical objections had not been overcome what effect would this have had on the development of science? What do your examples tell you about the way values influence the scientists' selection of a 'significant' problem in 'pure' science?

B.3 Write an essay putting yourself in the position of a member of another species advocating a world policy in which your interests are paramount.

C. The overall view: science as an activity

Is it possible to distinguish the abstract concept of 'science' from the practical manifestation of that concept? Today science is a massive and complex system of human activity and to complete our consideration of the issues involved in the controversy about the neutrality of science it is necessary to consider them in the context of this overall activity. At this level we are mainly concerned with the implications of the neutrality issue for another issue, that of whether or not the scientist has any social responsibility for the application of his work. This second issue is one which causes much heated debate between scientists. This divergence of opinion does not stop at debate, it leads to completely different views about how scientists should behave (viz. the formation of organizations such as the Union of Concerned Scientists and the British Society for Social Responsibility in Science). Underlying arguments about social responsibility, then, are often assumptions about the nature of science. Different assumptions lead to different and often conflicting conclusions. We can see this happening if we examine the papers by Black and Chain.

Black claims that, today, science as an overall activity can no longer be considered as the disinterested pursuit of truth. Even where scientists are working on the purest science which has no apparent practical applications scientists cannot escape the dilemma of responsibility because the speed of development is such that discoveries are often harnessed very quickly to industrial, military or other practical uses. Today this type of pure research is rare. Much research is aimed directly at specific objectives. It can no longer be considered neutral and is carried out with a definite purpose in mind: to increase the profits of industry or strengthen the power of government. Scientists involved in such projects know this and because the science is no longer neutral they have forfeited any claim to moral neutrality; they know what the intended purpose of their work is and they cannot subsequently plead 'not guilty' when this purpose is achieved and horror (or praise) is expressed at the results. Furthermore the way work is organized today means that scientists are often working in multi-disciplinary or vertically integrated teams so that basic research is being carried out at the same time as others are developing and applying the results to a specific objective (consider for example the work in plasma physics being carried out specifically with the aim of generating electricity from nuclear fusion). In such circumstances there is no realistic way of separating basic research from its application and, for Black, this harnessing of science to specific ends implies the end of scientific neutrality and with it the end of any legitimate claim to moral neutrality.

Chain takes a different view. He feels that the distinction between pure and applied science can still be made. Pure science is neutral in his view and hence scientists working in this field need have no moral qualms about possible applications of their discoveries. He then goes on to argue that scientists also do not have any responsibility either for harmful side-effects of the products of science or for the destructive consequences of using weapons they have helped to develop. In his view scientists are obliged to work directly towards fostering the aims of the political or industrial sectors: "Capable scientists are, therefore, the most precious asset which a nation possesses to give superiority over its enemies and victory or defeat is in their hands" . . . and again "The first responsibility of the scientist is to the nation of which he is a member". These conclusions are based on the assumption that science, whether pure or applied, is a morally neutral activity; it does not involve value judgements about what is acceptable or desirable. Such judgements are made by society as a whole and scientists, as scientists, have no special role to play on those decisions and no choice but to work towards achieving the society's objectives. Hence in talking of scientists working on the development of weapons he states that they do have a responsibility to explain the consequences of using such weapons, but only after these weapons have been developed. It would be immoral of them to stand back and refuse to apply their skills and knowledge to such ends. It seems that Chain views scientists as machines to be plugged in to perform whatever task lies within their capabilities. (Long ago Galileo warned that humanity was in danger of reaching this unhappy era: ". . . As things now stand, the best we can hope for is a race of inventive dwarfs who can be hired for anything".) While scientists retain the same rights as other citizens it is not for an individual scientist (or a group of scientists acting in concert) to challenge the aims of the state and of industry. Chain maintains that these are democratically established and must therefore be actively and loyally supported by all citizens (including scientists) until they are changed by 'normal democratic processes'. In other words he excludes any form of protest by a scientist qua scientist outside 'normal democratic processes'. Our purpose is not to examine the validity of these arguments, but to point out that one of the assumptions on which they rest is that science is a morally neutral activity, and that because of this the question of blame does not arise. Apparently the search for scientific knowledge and the development of its applications are neutral, only the decision to use the application is not so.

Discussion/essay suggestions

C.1 Should scientists take an active role in deciding how the results of their research should be applied? If so, what actions are open to scientists, either as individuals or as groups, who wish to prevent or promote specific applications of scientific knowledge?

C.2 If science is *not* neutral how is the claim that scientists should be free to study any natural phenomenon, regardless of implications, in the name of the harmless pursuit of knowledge, affected? If you conclude that some areas of research should be banned, what fields would you like to restrict and what mechanisms would you propose to implement such controls?

C.3 Our considerations of the various aspects of the neutrality of science have mainly centered on natural science. How far do the arguments we have discussed apply to the social sciences? Can social science be regarded as neutral in any sense?

See:

1.7 Easlea, B. (1973). *Liberation and the Aims of Science,* Chapter 6 (Chatto and Windus, and Rowman and Littlefield).

Further reading

All these readings contain assumptions, sometimes clearly stated but sometimes implicit, about whether or not science is neutral. In studying them students should look particularly for such assumptions and try to analyze how different ones about the nature of science would affect the conclusions of the various authors.

1.8 Kuhn, T. S. (1972). 'Scientific Paradigms'. In *The Sociology of Science.* Edited by Barry Barnes. London, Penguin, Chapter 4.
In this article Kuhn discusses the role of the paradigm in scientific research, illustrating that research work tends to be directed towards finding evidence to support the paradigm and demands a commitment from the scientist to the paradigm. This calls into question the image of the scientist as the objective searcher for truth.

1.9 Ezrahi, Y. (1972). 'The Political Resources of Science'. In *The Sociology of Science.* Edited by Barry Barnes. London, Penguin, Chapter 12.
This article examines how 'images' of different scientific disciplines have been used by scientists in the political battles which are fought over the allocation of resources, both between science and other activities, and within different scientific disciplines.

1.10 Pere, M. L. (1970). 'The New Critics in American Science'. *New Scientist,* 9 April.
The different ways in which scientists are becoming involved in political decisions in the USA are illustrated.

1.11 Anon. (1973). 'Building a Better Thumbscrew'. *New Scientist,* 19 July.
This discusses the role of scientists in promoting torture and condemns the lack of response from scientific communities to such situations.

1.12 Monod, J. (1971). 'On the Logical Relationship between Knowledge and Value'. In *The Social Impact of Modern Biology.* Edited by Watson Fuller. London, Routledge and Kegan Paul, Paper Two.
Discusses the role of science and scientists in a world where the discoveries of science have destroyed beliefs on which value systems were based and argues for a new value system based on "the ethics of knowledge".

1.13 Beckwith, J. (1971). 'The Scientist in Opposition in the United States'. In *The Social Impact of Modern Biology.* Edited by Watson Fuller. London, Routledge and Kegan Paul, Paper Eighteen.
Outlines the view that science is the basis of much exploitation and oppression and argues that scientists who believe this should adopt a radical role. Various ways that this can be done are discussed.

References for Chapter One

1. Chain, Sir Ernest (1970). 'Social Responsibility and the Scientist'. *New Scientist,* 22 October.
2. Galileo, G. (1953). *Dialogue on the Great World System.* Salisbury Translation. Edited by G. de Santillana. University of Chicago Press.
3. Bronowski, J. (1971). 'The Disestablishment of Science'. In *The Social Impact of Modern Biology.* Edited by Watson Fuller. London, Routledge and Kegan Paul.
4. Rose, S. and Rose, H. (1971). 'The Myth of the Neutrality of Science'. In *The Social Impact of Modern Biology.* Edited by Watson Fuller. London, Routledge and Kegan Paul.
5. Young, R. M. (1971). 'Evolutionary Biology and Ideology'. In *The Social Impact of Modern Biology.* Edited by Watson Fuller. London, Routledge and Kegan Paul.
6. See, for example, Caspari, D. W. and Marshak, R. E. (1965). 'The Rise and Fall of Lysenko'. *Science,* 16 July.
7. See, for example, Ezrahi, Yaron (1972). 'The Political Resources of Science'. In *The Sociology of Science.* Edited by Barry Barnes. London, Penguin.
8. Black, M. (1975). *Is Scientific Neutrality a Myth?* A lecture delivered to the Annual General Meeting of the American Association for the Advancement of Science, 27 January.
9. For a report on the present state of affairs in research on animals see Kendall, E. (1976). 'Should these animals die?' *Observer Colour Supplement,* 13 June.
10. Faulkner, W. (1962). *The Reivers,* (Random House, New York) or (Chatto and Windus, London, 1962) or (Penguin Books, Harmondsworth, Middx., 1976).

Chapter Two
Technology

ESSENTIAL READING

2.1 Dickson, D. (1974). *Alternative Technology and the Politics of Technical Change.* London, Fontana. Chapters 1, 2, 3, 7.
2.2 Eshelman, R. 'The Auto Safety Furore: Its Meaning to Engineering'. In *Engineer* July–August 1967; or In *Technology and Society.* Edited by N. de Nevers. New York, Addison-Wesley, 1972.
2.3 Hardin, G. (1972). *Exploring New Ethics for Survival: The Voyage of the Spaceship Beagle.* Viking. Chapter 7 (reprinted as Appendix 2).
2.4 Rickover, H. G. 'A Humanistic Technology'. In *Nature* **208,** 20 November 1965, or In *Technology and Society.* Edited by N. de Nevers. New York, Addison-Wesley, 1972.

A. Introduction

There is a very close relationship between science and technology and, in some instances at least, technology could be viewed as the direct application of science. In the eyes of the layman there may often be no distinction between the two. Consequently it is possible that views about the nature of science have been automatically transferred to technology without any serious thought about their appropriateness. As we have already seen there is considerable debate about the neutrality of science. In this chapter we look at the way similar conflicts have arisen about technology.

It is obviously inappropriate to try to use the word 'neutral' to describe technology in some of the senses that it was used to describe science in the first chapter. There is no way in which we can talk about 'the pursuit of knowledge for its own sake' or 'the objectivity of observations, experiments and theory' as applied to technology; technology necessarily implies the application of science, invention and industry/commerce to matters which are of importance to our life style and *must* therefore have a social effect. The exact nature of the relationship between technology and society is the subject of much controversy and discussion and the various theories are well summarized in the Open University Unit 2–3, *Technology and Society* by R. Roy and N. Cross, pp. 18–26, of their course Man-made Futures; Design and Technology.

But technology is undoubtedly commonly regarded as being neutral in some senses of the word. In this chapter we look at some of the ways that this happens. Whether technology can be considered neutral, in any sense, has important implications for the whole world because of the ever-increasing role which technology plays in people's lives.

B. Technology as a neutral tool

There is a commonly held view that technology, considered as a collection of machines, techniques and tools, is neutral in the sense that in itself it does not incorporate or imply any political or social values. Closely associated with this is the idea that technology itself is neither good nor evil. Any beneficial, or harmful, effects arise out of the motives of the people applying a particular piece of technology and the end to which it is used. Where a particular application, chosen for its beneficial results, produces harmful side-effects these are blamed either on inadequate social policies or on lack of sophistication in the control of the effects of technology. Whichever is chosen as the whipping boy, the technology itself remains 'neutral' (or blameless).

We can use industrial pollution as an example to illustrate these ideas. The solution to pollution is normally seen in two ways: *either* to limit pollution by increasing social control and sanctions, *or* to improve the techniques of pollution control. The technology which causes pollution remains unchallenged: a neutral tool chosen for technical, objective reasons as the most appropriate means of producing a specific end product.

The progress of technological innovation is seen as inevitable and unrelated to any considerations outside the state of scientific or technical knowledge. For example, Ferry says: "Technology has a career of its own, so far not much subject to the political guidance and restraints imposed on other enormously powerful institutions"[1]. Developments take place as a result of what is technically possible and objectively necessary. A particular choice about which piece of hardware to use may well be motivated by political, economic or social objectives, but the range of hardware which is available has been determined by technical considerations alone. The opponents of this idea of technology claim that it has arisen because certain characteristics associated with science have, unjustifiably in their view, been transferred to technology. Rickover explains this. He says that scientific knowledge, by its very nature, is true and pays no regard to whether people like the truth or to whether they feel it is acceptable. This need to disregard human considerations and the idea of objective truth have been transferred to technology. He says "A certain ruthlessness has been encouraged by the mistaken belief that to disregard human considerations is as necessary in technology as it is in science"[2]. He is, in fact, arguing that technology should not be 'neutral' in this sense, and that it should, in its development, take into account human needs and values.

In contrast to this Dickson takes the view that technology is not neutral. It already incorporates specific values and its 'inhuman' nature arises out of these values. He claims that the idea that technology is neutral is a myth which disguises the political role of technology. What is wrong is not that technology ignores human values but that the values that are already there are wrong and it is these which should be changed. *Scientism* ("an apparently 'scientific' approach to any problem or situation is both necessary and sufficient to indicate how its objective, politically-neutral resolution can be achieved"[3]) is blamed for being responsible for the generation of this myth. His ideas about the political nature of technology are discussed more fully in the next section.

There is a further sense in which technology is considered a neutral tool. This is in the sense of interchangeability or universality. It is considered to be neutral (in the same way as scientific knowledge) in the sense of being independent of a particular political or social system. Hence a successful product or process can safely be transferred from one country to another with complete predictability, just as, for example, a gas can be relied upon to conform with Boyle's Law whether it is in England or Egypt. Experience has shown that this is *not* necessarily, or even usually, the case. Consider, for example, the introduction of dried milk into Third World Countries.

Dried milk is an example of an apparently successful product of technology which is widely used and encouraged in the developed world. What happens, however, when this acclaimed achievement is transferred elsewhere? One account of the result says:

> Ignorant and uninformed adoption of European-type baby bottle-feeding also contributes to infant malnutrition in the towns. The practice of feeding powdered milk to babies is rapidly growing in urban areas, but this practice is being used by women with no knowledge of hygiene, no ability to read the instructions on the can, and no money with which to buy sufficient powdered milk. Thus, diluted powdered milk from dirty bottles and dirty teats is substituted for breast milk. This leads to malnutrition and dietary disorders such as marasmus, diarrhoea or vomiting . . . In Uganda, bottle-fed children were two pounds lighter on average.[4]

This is not the only problem which has been reported arising from the introduction of powdered milk. Lactose intolerance, leading to malnutrition, and vitamin 'A' deficiencies are widely reported following the supply of powdered milk by relief agencies. It is clear that the success of powdered milk in the developed world relies on social and cultural factors, the level of education and dietary conditions; many of these are completely different in the Third World situation. The successful transfer of other technologies will similarly depend on such factors which are frequently ignored in assessing the 'technical' solution.

Discussion/essay suggestions

B.1 Consider the present efforts to develop supersonic passenger transport in the form of Concorde. How far do you consider this an objective, neutral response to a perceived need, for the benefit of mankind, and how far is it a political maneuver?

See:

2.5 Braun, E., Collingridge, D. and Hinton, K. (1979). *Assessment of Technological Decisions: Case Studies.* Siscon Series. London, Butterworth.

B.2 What examples can you think of where technologists have implemented innovations without any regard for human considerations? Or is it more likely that the human considerations which have been taken into account are considered inadequate or inappropriate by the critics?

B.3 Harmful side-effects of technological innovation are seen as being overcome by the application of a more sophisticated technology; there is a technical solution to all social problems. Is this realistic or can you think of major social problems of today's world for which you feel there is no technical solution?

See:

2.6 Hardin, G. 'The Tragedy of the Commons'. *Science,* **162,** 13 December 1968, or In *Technology and Society.* Edited by N. de Nevers. New York, Addison-Wesley, 1972.

2.7 Weinberg, A. M. (1966). 'Can Technology Replace Social Engineering?'. *Bulletin of the Atomic Scientists,* 22 December, or In *Technology and Society.* Edited by N. de Nevers. New York, Addison-Wesley, 1972.

C. The political nature of technology

There are many different theories which try to relate changes in technology to changes in society. Some of these see technology as given, the independent variable which determines other facets of society.

Technology, determined largely by the state of scientific knowledge, controls what the social and political systems will be, what products are available, what types of work organization should be adopted. These decisions are thus *removed* from the arena of *political* and *social* debate and are claimed to be inevitable, the result of the imperatives of technology. This is further justified by assumptions that development is synonymous with economic growth which itself derives from industrialization and its associated technologies. It is in this rather different sense that technology is regarded as being *politically neutral.* The sense is no longer one of *not* influencing political and social systems but is now that of being *inevitable* if 'progress' is desired. The form that a particular technology will take is determined by objective, apolitical means such as technical feasibility and economic viability and the consequences which follow (for example, in terms of work organization and the means of control of workers) are determined by the technology alone. Any society which wishes to progress is bound to accept these consequences, whatever the values of the society.

Dickson does not agree with this viewpoint. He would accept that technology does lead to certain social, political and cultural ends, but he argues that technology is itself determined by political forces and is not a given, neutral, factor determined by the state of scientific knowledge. The social, political and cultural ends to which technology leads are the objectives of the controlling élite of society who so design technology as to make them inevitable. Both the nature of present technology and technological innovation are politically determined. Hence he states:

> In general we can say that a society's technology, when viewed as a social institution rather than a heterogeneous collection of machines and tools, is structured in such a way that it coincides with its dominant modes of action and interaction . . . Technology does not just provide, in its individual machines, the physical means by which a society supports and promotes its power structure, it also reflects, as a social institution this social structure in its design. A society's technology can never be isolated from its power structure, *and technology can thus never be considered politically neutral.*[3]

In *Alternative Technology and the Politics of Technical Change,* he points to several ways in which he believes this to be true. He claims that the nature of a society's technology at any given moment reflects and supports the dominant social class and its values. Technical developments or innovation are chosen not on the basis of objective necessity, but so that they will assist and support a particular section of society.

In the developed world technology has been applied to all aspects of life. We are dependent on machines and gadgets, not to mention the whole infra-structure of society (including the provision of food, power, water, sewage disposal, and transport and communication networks), to carry out many everyday activities. Some commentators in fact feel that we are dominated by technology and that the machine has taken over. Dickson argues that this has arisen because of the political nature of technology. This dominating technology reflects the wishes of the ruling class to control their fellow men. He uses the example of the methods of industrial production to illustrate this idea. He outlines the history of the Industrial Revolution and claims that the very process of industrialization did not arise from an objective assessment of production needs determined by economic factors. It arose from the desires of the dominant social class, the providers of capital, to dominate and control both Nature and work force. Consequently this set of values and desires was built into the design of the machines and factories. This process is still going on today he claims, and cites the introduction of containerization in ports and pre-fabrication of box-girder bridges as examples to illustrate how technological changes are designed to increase management control over the work-force, and asserts that it is disguised by statements about the objective necessity of the process for increased productivity and efficiency.

Many examples are given from domestic technology, the range of products and services which are available to the consumer, to illustrate that consumer choice is severely restricted by the very nature of mass production technology. The pattern of use is determined by technology, which is not a neutral tool to be applied to producing what the consumer wants. For example Dickson says:*

> The important question to ask about television, as of any major technological innovation, is not how it could be ideally used, but why it is used in the way that it is, and how does this usage become reflected in the design of the technology of communication as a social institution?

Dickson also illustrates a different aspect of the non-neutrality of technology by suggesting that its benefits are often confined to the members of a particular social class. He uses the development of the railway system in the United Kingdom†, with its emphasis on developing fast inter-city links and the pruning back of non-economic services to illustrate this:

> . . . The railway system can thus be seen as an 'objectification' of one of the modes of transport required by a capitalist economy, one that requires the rapid transit of its business executives –

* p. 89, op. cit.
† p. 178, op. cit.

and occasionally other classes, as in the case of migrant workers or those on holiday – yet is frequently able to ignore the interests of those less essential to the functioning of the economy, and indeed often able to withdraw services to them in the name of 'rationalization'.

Other societies, which started out with different political or social aims, made the mistake of copying the capitalist designed technologies, with the result that their societies in many respects took on the mantle of capitalism. They did this because they believed that technology was neutral and they could retain their own cultures and political values and systems whilst gaining the benefits which flowed from industrialization (capitalist style) and its associated technologies. They were not aware of the in-built values which meant that if they were to succeed in applying this technology they had to accept the values of the capitalist countries. Other technologies, designed to incorporate their own values, should have been developed if they wanted to preserve their own cultural and political systems.

Dickson argues that the generally accepted concept of technology as politically neutral is positively harmful in the consequences which follow from it. He claims it enables participation in the discussion about technical innovation to be denied to the vast majority, the people who are often most affected by the decision. The 'experts' who understand the 'facts' and who can be relied upon to make the 'correct' decision based on an 'objective' evaluation of 'economic necessity' are the ones who must debate.

Discussion/essay suggestions

C.1 Dickson uses the example of the development of the railway system in the U.K. to illustrate his ideas of how technology incorporates the value and requirements of the dominant system. Is it possible to think of alternative, technically feasible, ways in which a railway system could have developed? If so, what values, if any, would be incorporated within such systems?

Other examples with which you are familiar from your own technical studies could be used in the same way.

C.2 Consider the state of technological development in another country, either now or previously, with an ideologically different political system. Can you trace any differences, or similarities, between their technologies and the U.K.'s or the U.S.A.'s which can be attributed to *political* rather than *technical* or *economic* factors?

See:

2.8 Dean, G. (1972). 'China's Technological Development'. *New Scientist,* 18 May, pp. 371–373.

2.9 Sigurdson, J. (1973). 'The Suitability of Technology in Contemporary China'. *Impact of Science on Society,* XXIII, October–December, pp. 341–352.

D. Technologists at work

Whilst the work which scientists do varies considerably along the spectrum from pure research to applied technology, that of technologists is concerned almost exclusively with developing and implementing specific ideas with a definite end in mind. We are interested in this section in discussing how far the technologist is justified in considering himself a neutral cog in the industrial or military machine where he works. Is he there merely to perform a specific function with no consideration of the implications of that function, or does he have a responsibility to relinquish that apparent neutrality and take sides in any controversy over whether the end product is desirable or acceptable?

The debate can be considered at two levels. Firstly there is the question of the intended product of his work. How far, if at all, should the technologist make judgements about the desirability or otherwise of the end product in considering whether or not to apply his technical skill to a particular project? The second level arises from the unpredicted harmful or undesirable consequences which often arise from the application of a particular technology. How far is the technologist 'innocent' of responsibility for such consequences?

Eshelman[5] uses the example of automotive engineers involved in designing motor cars to illustrate the apparent lack of concern shown by engineers for the social impact of their products. He lists road deaths and injuries, high noise levels, congestion, pollution and despoliation of the countryside as some of the harmful consequences of motor cars and asks why the engineers and their associations have been silent about these aspects and continue to support management policies which place a low priority on such considerations, and, for example, put styling and performance before safety. This article assumes that engineers, if truly professional, have a responsibility to relinquish their neutral role and to take steps to limit the harmful consequences of their work. It discusses some of the factors which inhibit them from doing this and thus calls into question their status as professionals.

The opposite view, that technologists are not in any way responsible for the intended consequences of the use of their products, rests on the distinction between responsibility for the development of a product and responsibility for the use of that product. "I distinguish between developing a munition of some kind and using it"[6]. How far this is a real distinction is very much a question of individual conscience. It is also a matter of general concern as to whether society as a whole accepts this distinction and allows technologists, possibly working on developments with extremely evil consequences, to escape condemnation by sheltering behind this facade, the validity of which is at least open to debate.

We return now to the question of unforseen consequences of technological innovation and development. In the past there was a very limited understanding of the complex relationships between natural things. A machine or technique was developed for a specific purpose. The more responsible technologists may have considered its implications for a limited range of obviously related aspects and possibly made adaptations to restrict or remove the harmful consequences which they identified. Experience has shown however that this simply was not adequate. Repercussions of such developments have been far more widespread than even the most imaginative pessimist feared possible. In this context is it now necessary to impose a different philosophy on the control of technology? Is it still acceptable for the technologist to plead innocence when his device intended for human benefit turns out to do more harm than good? Hardin[7] thinks not. He introduces the concept of guilty until proven innocent and suggests this should be applied to all technical developments. (In one instance in the United States this principle has already been accepted: the burden of proof of both the effectiveness *and harmlessness* of medical remedies has been placed on the proponent.) The example of the development of detergents and the unforeseen snags along the way is used to illustrate the difficulty of trying to guard against unexpected consequences. This view of the technologist as being responsible for all the effects of his innovation denies him any shelter behind the neutrality shield. Technology and, by implication, the people working with it, is no longer a neutral tool which can be used for good or evil, but must in the first instance, according to Hardin's proposal, be *assumed* to be harmful until proved otherwise. He accepts that technology *per se* has both beneficial and harmful potential, it is neither good nor evil. He is, however, proposing a practical procedure for reducing the harmful consequences of introducing new technologies.

Discussion/essay suggestions

D.1 Scientists and technologists are frequently attacked for refusing to take sides on controversial issues involving the application of scientific knowledge and producing the goods whatever the implications. Is there any reason why they should be singled out for attack when all employees are in fact contributing directly or indirectly to the same end product? Is the technologist who designs and develops a weapon, for example, any more responsible for its use than the craftsman who assists in manufacturing it or the clerk who keeps production records? If so, what factors distinguish his role from the others in this context?

D.2 If a country was to accept the concept of 'guilty until proven innocent' it would have to define the criteria which would be accepted as proof of innocence. Make a list of the criteria you consider appropriate to meet this objective and then consider a number of recent technological innovations to see how they measure up to your standards.

Further reading

2.10 Ferry, W. H. 'Must we Re-write the Constitution to Control Technology?'. In *Saturday Review,* **51,** 2 March 1968, or In *Technology and Society.* Edited by N. de Nevers. New York, Addison-Wesley, 1972.
Argues that uncontrolled technology threatens the social and personal welfare of Americans and that the days of *laissez faire* technology should be ended with far-reaching legal changes designed to limit the extent and direction of technological change.

2.11 Roy, R. and Cross, N. (1975). *Technology and Society,* Units 2 and 3 of the Open University course, "Man-Made Futures: Design and Technology", T. 262, particularly pp. 18–26.
These units summarize the many theories which relate social and technological change and who controls technological change. The whole content is relevant to the problems we have discussed, but Section 2 is especially useful as it summarizes the models which have been devised to try to explain the relationship between technological and social change.

2.12 Ferkiss, V. (1969). *Technological Man: The Myth and the Reality.* London, Heinnemann.
The whole book is concerned with illustrating his central thesis that man's latest technological capability and potential (genetic engineering, for example) raise the possibility of fundamental changes in human life and that man must change his life style to take advantage of this situation. Changes in economics, politics, values and cultures are necessary. A sampling of any chapter will give students an appreciation of his ideas.

2.13 Mathes, J. C. and Gray, D. H. (1975). 'The Engineer as a Social Radical'. *Ecologist,* **5,** May.
Because of the major impact that technology has on society, technologists, in spite of their self-image and that of others, are the new radicals whose activities affect such things as the distribution of power and decision-making in society, personal-community relationships and family structures. If technology is to be controlled, the engineer will have to reconcile this conflict.

2.14 Pelto, P. J. (1973). *The Snowmobile Revolution: Technology and Social Change in the Arctic.* Cummings.
A study of the repercussions on the lives of the Skolt Lapps in N.E. Finnish Lapland on the introduction of Snowmobiles. It describes the changes which are taking place in the economic system, in transportation and in patterns of social interaction as a direct result of the new technology.

References for Chapter Two

1. Ferry, W. H. (1968). 'Must we Re-write the Constitution to Control Technology?' *Saturday Review,* **51,** 2 March.*
2. Vice Admiral Rickover, H. G. (1965). 'A Humanistic Technology'. *Nature,* **280,** 20 November.*
3. Dickson, D. (1974). *Alternative Technology and the Politics of Technical Change.* London, Fontana.
4. Hughes, C. C. and Hunter, J. M. (1973). 'The Role of Technological Development in Promoting Disease in Africa'. In *The Careless Technology.* Edited by T. Farvar and J. P. Milton (Tom Stacey, Prospect for Man series).
5. Eshelman, R. (1967). 'The Auto Safety Furore: Its Meaning to Engineering'. *Engineer,* July–August.*
6. Quoted in Black, Max, 'Is Scientific Neutrality a Myth?'. Appendix One to this book.
7. Hardin, G. (1972). *Exploring New Ethics for Survival – The Voyage of the Spaceship Beagle.* Viking Press, Chapter 7.

* Reprinted In *Technology and Society.* Edited by N. de Nevers. New York, Addison-Wesley, 1972.

Chapter Three
Intermediate Technology

ESSENTIAL READING

3.1 Schumacher, E. F. (1974). *Small is Beautiful.* Abacus, Part III, Chapters 11–14.
3.2 Dickson, D. (1974). *Alternative Technology.* London, Fontana, Chapter 6.
3.3 Reddy, A. K. N. (1975). 'Alternative Technology – A Viewpoint from India'. *Social Studies of Science,* **5**, pp. 331–342.

A. What is intermediate technology?

Read again Section A of Chapter Two. What is said there still applies.

The term 'intermediate technology' was first used by Schumacher to describe the technology which would provide jobs (called workplaces by Schumacher) for an amount of capital investment between the very small sum required in traditional societies and the very large one required in the complex technology of the industrialized world. He symbolizes his ideas by expressing traditional technologies in developing countries as £1-technologies and those of developed countries as £1000-technologies. This difference represents a vast gap which is impossible for most developing countries to cross. He proposes an intermediate level of investment, symbolically £100, which would allow large numbers of jobs to be created quickly. It would also rapidly increase productivity and provide businesses which not only suited the financial conditions, but also fitted in to the existing levels of education, skills, business experience and know-how. This is just one definition of a whole range of ideas which come under the general term 'alternative technologies'. Our discussions in this section will be limited to discussing this particular definition and the implications which follow from it.

He identifies four criteria which must be met by such a technology. These are:

> First, that workplaces have to be created in the areas where the people are living now, and not primarily in metropolitan areas into which they tend to migrate. Second, that these workplaces must be, on average, cheap enough so that they can be created in large numbers without this calling for an unattainable level of capital formation and imports. Third, that the production methods employed must be relatively simple, so that the demands for high skills are minimized, not only in the production process

itself, but also in matters of organization, raw material supply, financing, marketing and so forth. Fourth, that production should be mainly from local materials and mainly for local use[1].

Hence intermediate technology does not refer to any particular industry or technique. It is an all-inclusive term for any industrial process which fits within these categories. 'Industrial' is also intended to include technologies which can be applied to improving agricultural industries. One of the justifications frequently used for emphasizing the importance of intermediate technology is that it has low impact. It is claimed that industrialization via intermediate technology has little or no effect on societies. The small disturbances that might arise can be absorbed by the society without necessitating major changes in the traditional way of life. Within this idea is an implication of neutrality.

B. Aims

There is a realization that previous development strategies in the Third World have not been successful. In fact some people argue that the gaps between developed and developing countries and between the rich and the poor within particular Third World countries are widening. Two consequences, in particular, are causing concern. These are the growth in unemployment in rural areas and the drift to the towns by the rural population in search of job opportunities. Intermediate technology is seen as the solution to these two problems. It aims to provide large numbers of jobs in rural areas, thus, at one and the same time, reducing rural unemployment and encouraging people to stay in the villages.

The existence of two sectors, a traditional rurally-based society and a modern one based on towns and incorporating modern technological development, is accepted. Intermediate technology is intended to tackle the problems within the rural sector and ignores the development of the modern sector, which, according to Schumacher will inevitably receive attention and finance through various existing mechanisms. Hence intermediate technology can be seen as a technical solution to a perceived social problem. The problems have been identified by observing the situation in the Third World, they have been attributed to the wrong choice of technology (i.e. the transfer of complex, capital intensive, modern technology) and the solution is seen to lie in correcting this error (i.e. the choice of a more appropriate technology). Thus a social problem is to be resolved by a 'neutral' act, political and social changes not being required, to make the society more 'efficient' in achieving development.

C. Is intermediate technology neutral?

If intermediate technology is to be considered neutral, it must meet certain criteria. From the ideas discussed in Chapter 2, we can draw up a list of such criteria.

1. It must not incorporate any political or social values. We can explain this a little more clearly by looking at what this implies. It means that this type of technology would be available to any society which had access to the necessary technical expertise, whatever their political or social goals (other than the stated goal of full employment in a rurally-based society). It also means that by adopting such a technology the people (or government) are not committing themselves to any particular social or political system, and that adoption of such a technology would not predetermine what alternative types of organization were available to that society. Questions such as in whose hands power should lie, how wealth and ownership should be distributed, or what social services should be provided must not be influenced or predetermined by the adoption of intermediate technology.
2. The choice of intermediate technology must be an 'objective' decision (made on the basis of technical feasibility and economic viability) that this is the most appropriate technology to solve the particular problem in mind.
3. There will be no further effects, whether good or bad, except the stated goals, i.e. full employment in a rurally-based society.
4. Technologists working in this field are concerned only with devising particular 'bits of technology' to meet the criteria of intermediate technology and fit the particular task in hand. Their role is one of the passive engineer carrying out a technical task, e.g. devising the equipment necessary for harnessing solar energy on a small scale. As *engineers* they have no responsibility for or concern about any wider ramifications of their work as long as they succeed in creating the technology which will provide jobs of a 'suitable' type in the rural situation.

Now let us consider each of these criteria and examine how intermediate technology meets the requirements. We can try to assess this in several ways. We can examine the definition of intermediate technology to see what, if any, political or social values, are incorporated into this definition. We shall analyze statements about intermediate technology by technologists working in the field and also investigate some examples either to see what has resulted from the application of intermediate technology or to make predictions of possible consequences.

There are many issues arising out of the application of intermediate technology and of development strategy in the Third World. These are far too complex and wide ranging to be discussed here. We are not concerned in this book with trying to decide whether or not development will follow from the widespread application of intermediate technology. The whole question of whether economic development is either desirable or desired is also outside our terms of reference. Our only purpose is to consider whether intermediate technology, as defined by its proponents, can be considered neutral in any way.

D. Philosophy and definition

Now let us consider the definition and philosophy of intermediate technology as given by Schumacher.

The definition of intermediate technology embodies a number of value judgements. Examples are:

1. that any work is better than no work, and the need to work is paramount in the hierarchy of human need: "For a poor man the chance to work is the greatest of all needs, and even poorly paid and relatively unproductive work is better than idleness."[2]
2. industrialization is a pre-requisite for development, which is seen only in terms of economic growth.

Whilst these arise from the definition of the problem which requires a solution, i.e. urban drift and rural unemployment, it is useful to see that these are embodied within the concept of intermediate technology. Rural life is to be made more attractive by the provision of jobs, which, in effect, kills two birds with one stone. The definition of the problem has been influenced by the Western concept of 'unemployment'. People observed to have no regular paid tasks are classed as 'unemployed' and the solution is seen to be 'provide employment'. In traditional societies, however, there was no such thing as 'unemployment'. People were engaged in subsistence farming or other tasks which contributed to making the society as self-sufficient as possible, and trading arrangements, usually independent of a monetary economy, existed with other societies to provide essential or prized items not available locally, e.g. salt. Tasks were distributed so that everyone had a role to play. If the problem had been defined in another way, for example: *How can we prevent the population from abandoning their traditional cultures, from being unduly influenced by the products and philosophy of the Western World and from losing their feelings of community responsibility*?, the solution may have been seen in an entirely different way. In such a case, the creation of job opportunities and paid employment might have assumed low priority. The important point here is that a technical solution to a social problem tends to build into itself the values which are held to be important by the way the problem is defined. If the same problem is defined in another way, different values are seen as being important and a different solution, embodying those values will follow. The way in which the problem is defined, and hence its perceived solution are a function of social, cultural and political factors.

A decentralized, regional approach to planning is seen as a prerequisite for the application of intermediate technology. This obviously has major political and social consequences. Intermediate technology appears to be dependent, if it is to be successful, on a specific type of political organization. In other words, intermediate technology is not a tool which could be applied successfully in any type of political system, and, presumably, if it were adopted by an unsuitably organized society, it would either fail, or force political and social changes upon that country.

One predicted result of the introduction of this type of technology is an increase in the number of people with entrepreneurial ability. This is again felt to be a necessary prerequisite for development. It is not absolutely clear how it is intended that this ability will be used but analysis of comments does suggest that small individually owned businesses will predominate, and from these large businesses with hierarchical management structures will develop. In this way the entrepreneurs will be able to extend their talents, through experience, to meet the growing demands from larger and more complex organizations. For example, Schumacher says:

> The introduction of an appropriate intermediate technology would not be likely to founder on any shortage of entrepreneurial ability. Nor would it diminish the supply of entrepreneurs for enterprises in the modern sector; on the contrary, by spreading familiarity with systematic technical modes of production over the entire population, it would undoubtedly help to increase the supply of the required talent.[3]

Dickson draws attention to this fact and suggests that intermediate technology is merely a way of encouraging a capitalist economy, and that this is the objective of the people who are promoting it. He argues that before intermediate technology can bring any real improvements there will have to be major political and social changes and that these must affect the modern sector. It is impossible, in his view, to treat the rural and modern sectors as being independent.

In other words technology is not a neutral tool, but part of the political process of control. Intermediate technology will also become part of the existing political system and not bring the positive results of development, unless it is accompanied by political and social changes. Dickson then goes on to analyze examples quoted by proponents of intermediate technology as illustrating its success and suggests that the success is, in effect, due to the political and economic systems of the countries concerned. He uses China, North Vietnam and Cuba as examples.

E. Motives of engineers

How do engineers who are working in this field view the role of intermediate technology? Do they believe that this is a technical solution to a social problem in developing countries and the problem will be resolved simply by changing the technological base of those societies? Or does it form part of a whole philosophy which suggests that this particular social problem is a product not *merely* of the wrong technology but also of the wrong social and political organization? In this case the adoption of intermediate technology is just one aspect of the necessary change and engineers who work in it and who hold this second view would see political and social change as being another side of the same coin.

Reading 3.3 is a statement by Reddy, an Indian scientist deeply committed to intermediate technology. This is essential reading for this chapter. In his paper the author extends the social problem which intermediate technology is intended to overcome far beyond the rural unemployment and urban drift specified by Schumacher. He includes the existence of an affluent powerful élite and great inequalities in Indian society, as well as chronic balance of payments difficulties, amongst the evils caused by Western technology. A change of technology will contribute to removing these ills, but *not* on its own. He clearly views his work in intermediate technology as part of the total process of bringing about a fundamental change in the nature of Indian society, a change which is technological, social and political. He warns of the dangers that intermediate technology can serve the rich and increase inequality. There is no doubt that neither Reddy, as a technologist at work, nor the technology which he advocates can be regarded as neutral when looked at through his eyes. He has concluded that technology is not a neutral tool which can be changed at will to solve social problems, but a value-laden political weapon which must be used in conjunction with appropriate political action to bring about the required change in society. Notice too that Reddy sees the choice of technology as a political decision, determined by those in power and not an objective 'scientific' decision.

Other views, too, have been expressed. Consider for example the following statement:

> As a developing country becomes more prosperous it generates a demand for capital goods – a demand which, at least in the initial stages, must be satisfied by exports from the developed world. This is already happening in many developing countries, where contrary to expectations, a gradual industrialization has led to great imports of capital goods from the industrialized nations.[4]

Here an engineer working in the U.K. on intermediate technology for developing countries explains that intermediate technology has brought benefits to the industrialized world by increasing markets for their products. His aim seems to be to create a materialistic society with demands which cannot be met from within so that his society will benefit. This is a long way from the original idea of removing the dual evils of rural unemployment and urban drift (or as we re-phrased it: 'preventing the population from abandoning their traditional cultures, from being unduly influenced by the products and philosophy of the Western world and from losing their feelings of communal responsibility'). What, then, are his motives and can he be regarded as neutral?

F. Specific examples and their implications for neutrality

We turn now to actual examples of intermediate technology and see how they measure up to our criteria of neutrality. Van Rensburg provides one example[5]. He wanted to introduce a number of small-scale, village-based breweries into Botswana. Lager beer is already imported from South Africa and there was a proposal, opposed by van Rensburg, to build a large modern brewery in the capital of Botswana. He advocated the application of intermediate technology by building several small-scale, village-based breweries to assist in meeting the two required objectives, i.e. creation of job opportunities at the village level and reducing urban drift. What possible effects could there be? One possibility is that distribution and supply of beer throughout the country would be made easier and hence make beer more readily accessible. This could have considerable social implications – for example decreasing standards of nutrition because of channelling of money from food to beer, increasing crime rate because of the effects of alcohol, and neglect of traditional work. Similar consequences have been observed in other developing countries where alcohol has become readily available. Should this be the case, then these consequences follow *directly* from the decision to have *village-based breweries*, i.e. the decision to *provide beer by an intermediate technology method.*

Another difficulty is illustrated by a further example. Less than 50 years ago, white men first entered the valleys in the Highlands of Papua New Guinea. They took with them metal axes as presents to create good-will amongst the indigenous people. Although this was not a deliberate attempt to promote economic development it can be viewed as an example of intermediate technology – metal axes replaced traditional stone axes used for chopping down trees, making gardens and other tasks associated with their largely self-sufficient subsistence economy. Relatively little investment was required and the metal axes were more productive than traditional methods. One estimate is that some tasks took only one-third as long using metal axes. No change was necessary in the organization or control of village life, no new products were introduced to disrupt the traditional society and no factories or alien methods of work were created. The result of this improved efficiency was *not* increased output or new products and services for the community. Salisbury summarizes the outcome:

> In the first place the introduction of the new steel technology raised the potential supply of goods of all kinds, since it set free time that could have been used to make any kind of good. No more subsistence goods were produced, since demand for them was stable given the existing role structure of society. Time was spent in efforts to increase the power of each individual and the group. Some of these efforts took the form of fighting to obtain power; some were efforts to obtain power through the increased use of valuables.[6]

This pattern has also been observed in other traditional societies. It seems that there are many other important factors besides technology which determine whether economic development will take place. It is true that the circumstances in this example are very different from those in some countries where intermediate technology is being advocated. Nevertheless, an important lesson may be that the consequences of improving the efficiency of traditional occupations are not necessarily the increased job opportunities predicted. It may serve merely to allow people already engaged in productive tasks to have more time for preferred non-productive (in an economic sense) occupations. Whilst this is not necessarily a bad thing it does suggest that, at least in some circumstances, intermediate technology will bring about changes different from those which it is designed to encourage. On the criterion of bringing about no changes other than the stated goals, this example of intermediate technology appears to fail.

G. Conclusion

We can now summarize the evidence of whether intermediate technology meets the requirements for neutrality as set out in Section C.

Firstly intermediate technology has been seen to demand a decentralized, regional approach and the encouragement of an entrepreneurial (capitalist?) approach to business. These are political goals which are *not* necessarily implied by the problem to be solved, but which, apparently, are built in to the idea of intermediate technology.

Technical feasibility and economic viability, although important, are not sufficient criteria for the adoption of intermediate technology for some people. According to Reddy a major criterion is the political one of whose interest it serves, i.e. it is not whether it provides jobs in the rural situation, the stated goal, which is important, but whether or not it will *reduce* the inequalities in Indian society. It is possible, of course, to provide jobs and maintain or even increase inequality. This is not good enough and so we have a clearly political criterion over-ruling technical and economic considerations.

We have seen, too, that at least for two engineers working in this field their motives can be seen as political. One is concerned with creating a new society in India and the other with encouraging patterns of life in developing countries which will assist the Western world by improving its markets. They both seem to be concerned with how technology is used, not just with producing the hardware. The examples which we have looked at also illustrate that good intention is not enough and that there will almost inevitably be other effects beyond the main objectives of intermediate technology.

Hence on all four criteria discussed earlier intermediate technology appears to fall down. Perhaps we have appeared to be stating the obvious in this section. 'Intermediate technology' is used in two senses: one to describe bits of technology: machines, tools etc; the other to describe a philosophy about the nature of society. The two aspects are closely linked with the philosophy being embraced by the technologist and being built into the hardware produced by them. An important question to ask is how far this is true of all technology and, in the final analysis, the science on which technology is based? Many exponents of intermediate technology are quite clear and quite open about their philosophical/political approach. It has arisen from a positive effort to analyze sophisticated Western technology and the impact it has made on non-Western societies. Perhaps the fact that Western technology has arisen out of Western society has obscured the real nature of our technology from us and created an illusion of neutrality. In the words of David Edge:

> AT (alternative technology) began with the realization that advanced, capital-intensive Western technology was 'non-neutral': but the alternative is *not* a 'neutral' technology – it is a technology *committed to different values*'. Every AT practitioner knows in his bones that technology is intimately value-loaded; he therefore specifies the values and social/political goals he has in mind, and designs technology to 'match'.[7]

Discussion/essay suggestions

A.1 We have concluded that Intermediate Technology is not neutral. How far is this conclusion related to the fact that Intermediate Technology is seen as a technical solution to a *social* problem, and that, because of this, it has social and political implications? Are there purely *technical* problems which can be solved by the application of technology and if so would analysis of such an example reveal a different answer to the neutrality question?

A.2 A group in India are working on improvements to the traditional bullock cart to increase its speed and travelling distance capacity. What possible unexpected effects can you predict from the widespread introduction of a more efficient, low-cost means of transport?

A.3 Aziz argues that one of the reasons for the apparent success of development policies in China is that Intermediate Technology has arisen out of the commune lifestyle. Developments of this nature in other countries are often imposed, either by central government or by aid officials from outside the local community. Is this difference in approach likely to have different effects and, if so, what are the implications of this for the neutrality argument?

See:

3.4 Aziz, S. (1975). 'China's New Economic Order'. *New Internationalist,* No. 32, October.

Further reading

It would be useful for students to have some understanding of the problems of development as background information for this chapter. There are innumerable publications in this field all of which should give an appreciation of at least some of the difficulties. The ones listed below should be readily available.

3.5 Hensman, C. R. (1971). *Rich against Poor: The Reality of Aid.* Allen Lane.

3.6 Donaldson, P. (1973). *Worlds Apart: The Economic Gulf between Nations.* London, Penguin.

3.7 Berstein, H. (ed.) (1973). *Underdevelopment and Development: The Third World Today.* London, Penguin.

3.8 Jale, P. (1968). *The Pillage of the Third World.* Monthly Review Press.

3.9 Hayter, T. (1974). *Aid as Imperialism.* London, Penguin.

3.10 *The New Internationalist* – a monthly magazine specializing in problems of world development. Issue No. 32, October 1975 is a special one featuring the New Economic Order.

3.11 Farvar, M. Taghi and Milton, J. P. (eds.) (1973). *The Careless Technology.* Tom Stacey.
This is the report of the proceedings of a conference sponsored by the Conservation Foundation and the Center for the Biology of Natural Systems, which by means of case studies illustrated the relationship between large-scale technological developments and ecological damage in the Third World.

3.12 O.E.C.D. Development Centre (1974). *Choice and Adaptation of Technology in Developing Countries — An Overview of Major Policy Issues.* OECD.
Contains a large number of papers relating to various issues arising out of the introduction of non-traditional technologies into the Third World. It includes a section on intermediate technology (Part 3, Section 1).

3.13 Kestenbaum, Ann (ed.) (1975). *Technology for Development.* VCOAD.
A collection of extracts highlighting issues arising out of the role of science and technology in Third World Development. Produced for use by teachers 'interested in bringing either a "technology element" into studies of development or a "Third World element" into studies of technology in society', it provides a useful introductory over-view of many aspects of development.

Intermediate Technology is just one example of a whole range of 'alternative' technologies. These are seen as answers to the present problems caused by the application of sophisticated technologies. Information about these would provide relevant background material. *See* for example:

3.14 Dickson, D. *Alternative Technology and the Politics of Technical Change.* Chapters 4 and 5.

3.15 Harper, P. 'Soft Technology'. In 3.12 above.

References for Chapter Three

1. Schumacher, E. F. (1974). *Small is Beautiful.* Abacus, p. 146.
2. Schumacher, E. F. (1974). *Small is Beautiful.* Abacus, p. 145.
3. Schumacher, E. F. (1974). *Small is Beautiful.* Abacus, p. 155.
4. Kenkare, A. S. (1975). 'Technology for the Developing World'. In *The Chartered Mechanical Engineer,* March.
5. *See* van Rensburg, P. (1975). 'Small and Simple Breweries for Rural Africa'. In *The Brewer,* March.
6. Salisbury, R. F. (1962). *From Stone to Steel: economic consequences of a technological change in New Guinea.* Melbourne University Press, p. 205.
7. Edge, D. O., personal communication.

Appendix One
Is Scientific Neutrality a Myth?

By Max Black

The question is not intended to be a rhetorical one. The complex and momentous issues connected with what is often called the "social responsibility of scientists" deserve the coolest and most sober consideration. My purpose today is not to moralize and not to preach. But I shall try to make a modest contribution towards unraveling a tangle of logical and moral issues that need to be considered.

The implications of "neutrality"

Let us begin with the notion of "scientific neutrality" and ask what it really means. Although it is widely used, the meaning of the expression is unclear and its suggestions can easily be misleading. The dictionary meaning of neutrality is that of a policy of withholding aid from any states at war with one another. More generally, neutrality is the attitude of one who takes neither side in a dispute – one who remains indifferent in the presence of some dispute or conflict.

In the context of "*scientific* neutrality", calling science "neutral" implies that scientists are *non-combatants.* But in what war, in what dispute, are scientists supposed to take no sides and to remain indifferent? (Incidentally, throughout this talk I shall be thinking of the obligations of scientists. If I speak of "science" that should be regarded only as convenient shorthand.)

The conflict in which science is supposed to stand aside can be conceived, for a start, as a conflict between *practical policies.* Consider the question whether to smoke cigarettes or not. The prevailing view is that science can only provide *factual* information that is relevant to this issue and has nothing to say about the *values* that are involved. Such factual information can be ammunition for either side, because it can strengthen the case either for smoking or for non-smoking. But it would be wrong, as many people think, for a scientist *as a scientist* to tell an individual smoker or the society to which he belongs which choice to make. Whether we choose to take the risk of lung cancer, or choose to ignore that risk for the sake of the pleasures of smoking, is up to you and me. On this view, science has, above all, nothing to say about whether lung cancer is an evil. That crucial judgement is also up to the consumer. On this view, expert scientists are supposed to say to their fellow citizens: "Whether you take lung cancer to be good or bad is your affair. And anything you propose to do about the risks and dangers of cigarette smoking is also your affair. As scientists, we have nothing to say about such value judgements or the appropriate policy decisions. We stand neutral." According to this view, a similar posture of "neutrality" is also in order – is indeed, required – in such more controversial issues as, say abortion or medical experiments upon uninformed subjects.

The conflict in which scientists are supposed to remain neutral may also involve a clash between moral positions rather than between practical policies. Consider, for example, the question whether a nation is justified in committing genocide. According to the prevailing view, the scientist, as scientist, has nothing to say about this. If a nation wishes to engage in genocide, science can provide the tools – the most efficient means for achieving that unchallenged end. But should a nation reject genocide as a monstrous crime, that too is its right and privilege. As between favoring genocide and opposing genocide, science is neutral: it has, one might say, *no moral dimensions.* The doctors who experimented on the hapless inmates of Nazi Germany's concentration camps cannot be reproached *as* scientists.

This conception of the scientific expert as one who stands aside from the clashes of policies and moralities – as, say, Switzerland stood aside from the war between Germany and the Western Powers – is, however, misleading in one important respect. A neutral is one who *chooses* not to take sides, and must therefore necessarily have the powers to be a partisan if he wished. But if "scientific neutrality" is a reality and *not* a myth, scientists *have no choice* about taking sides on the agonizing personal and social problems of our time. But then talk about the "neutrality" of science smacks somewhat of the ridiculous – like calling bread neutral because it can feed sinners or saints indifferently. There can be no *scientific* choice, no scientific partisanship in matters of policy or morality; and hence there can, strictly speaking, be no scientific neutrality either: science just has no *direct* bearing upon any except *scientific* issues. The qualification is, of course, important. The "neutrality" of science, one might say, has to be *essential*, incapable of removal or amelioration. But perhaps it would, on this view be better to speak of scientific *irrelevance.* Nevertheless, I will continue to use the convenient, if somewhat misleading label of "neutrality".

The thesis of the essential neutrality of science might usefully be broken down into two sub-views. If science really can say nothing about what *should* be done or should not be done, it might be called *normatively neutral.* If it can say nothing about what is good or bad, right or wrong, it might be called *evaluatively neutral.*

Whether in its normative or evaluative version the neutrality thesis depends upon the answers to some troublesome questions of *logic.* If science really is normatively and evaluatively neutral that is something we shall just have to accept: to be morally indignant about it would be as absurd as to complain that 2 and 2 make 4. For we have to deal in the first instance with a question of logic, not one of morality or policy.

The logical defence of the neutrality thesis

Let us now consider the reasons for thinking the neutrality thesis to be correct. So far as I know the only reason in its favor, but a powerful one, is a supposed proof of the existence of a "logical gap" between 'is' and 'ought', or between 'is' and 'should'. More specifically, there are supposed to be conclusive reasons for recognizing that no proposition containing words like 'should', 'should not', 'must', 'must not' – that is to say, normative words – can follow from the factual propositions that science seeks to certify as true or probable.

The classical form of this argument is due to David Hume and continues to be a part of the current orthodoxy of nearly all professional philosophers. Hume's argument is straightforward and deceptively simple. In a famous passage from his Treatise[1], which has been quoted a thousand times, he maintains – to use his own language – that all the moralists he knows make an imperceptible transition from observations about human affairs or assertions about the existence of God, all expressed with "the usual copulation of propositions, *is* and *is not*," to normative conclusions "connected with an *ought* or *ought not*." He says that this transition is "of the last consequence" and needs to be explained. In his own words: "It is necessary . . . that a reason should be given for what seems altogether inconceivable, how this new relation can be a deduction from others, which are entirely different from it." The nerve of Hume's argument, therefore, is the altogether plausible idea that the conclusion of a valid deductive argument cannot contain any material that is not already contained in the premises. If the premises are factual, and so contain no reference to *ought* or *ought not* (or for that matter to good or bad) it does seem "inconceivable," as Hume says, that normative or evaluative conclusions should follow.

Of course there is one way in which new material of a normative or evaluative sort *can* be present in a valid conclusion. From the premise that omelets cannot be made without breaking eggs, which can be regarded as "factual," it certainly follows that *if* omelets ought to be made *then* eggs ought to be broken. What Hume means is that no *categorical* or *unconditional* 'ought'-conclusions can follow from factual premises. This certainly seems, at least at first sight, extremely persuasive. But the matter is not so clear. How do we know that a given proposition is factual, non-normative, in the relevant sense? The absence of the crucial words, 'ought' or 'should' is not a reliable criterion. For example, the proposition that murder is a sin, certainly does logically imply the conclusion that one ought not to commit murder. But how are we to tell from the mere linguistic form of the first proposition that it is non-normative? Well, one might fall back upon the idea that the proposition about murder is *unverifiable.* But that raises further and controversial questions that are not part of Hume's formal argument. His famous argument needs to be supplemented by further considerations about what propositions shall count as being in some appropriate sense "factual" or "objective" – that is to say, true or false independently of the desires or hopes of the persons who assert and believe them. Behind *this* issue there lurk difficult questions about how we should conceive of the aims and procedures of scientific investigation. Should we, specifically conceive of science as a very special and distinctive way of approximating to knowledge – or as one sharing the goals of the whole spectrum of knowledge-seeking activities?

My own considered judgement is that Hume's argument in the form in which I have presented it, is circular. I believe that certain categorical 'should' – and 'ought' – propositions do have truth-value, and can be certified as true objectively (independently of desires or hopes). If so, Hume was wrong in thinking them "altogether different from scientific propositions. In a properly comprehensive sense of knowledge – which is broader than that of *scientific* knowledge – narrowly conceived, some normative and evaluative propositions can be known to be true. But if so, Hume's argument collapses and better arguments must be found for the logical segregation of scientific truths from practical and moral issues.

These questions are too intricate and controversial for detailed discussion here. I mention them in order to counteract the widespread opinion that somehow the independence of 'is' from 'ought' has been established once and for all by logicians and philosophers. To counteract this weak argument from philosophical authority, it is worth reporting that a sizeable number of contemporary philosophers have rejected Hume's argument.

But in a way it does not matter for our present purpose whether Hume's critics are right or wrong. I think we are entitled to assume that human beings, however diverse their moral or religious background, can agree upon certain fundamental ethical principles. We must postulate this in order to make any kind of rational *dispute* about means or ends possible. If we were to encounter somebody who thought it all right to eat young children in order to increase our food supply, argument would be pointless and therapy or something more drastic would make more sense: anybody genuinely holding such a position would have to be treated as monstrously inhuman or insane. So even if factual truths had to be regarded as logically segregated, the introduction of some generally acceptable normative conclusions would legitimate the derivation of normative conclusions.

More importantly, however, I shall argue later on that any approach to our problems that confines attention to the *products* of scientific activity, considered as isolated *propositions*, and any approach that conceives of knowledge as merely the justified *assertion* of such propositions is misleadingly and inappropriately abstract. The weightiest, most urgent, problems of "neutrality" are unaffected by the supposed logical gap between scientific truths and norms or values. If we think of "science" more concretely and realistically as something that human beings do, the issues, as I shall hope to show, look very different.

On the whole, the thesis of the gap between 'is' and 'ought' although it has had enormous appeal and very widespread currency, deserves in my opinion to be called one of the great half-truths or "popular fallacies" of contemporary Western culture. We should especially beware of the seductive but invalid syllogism that runs as follows: science is a good thing; science is necessarily neutral; therefore scientific neutrality is a good thing. For this syllogism might easily be converted, by negation of the conclusion, into: Scientific neutrality is a bad thing; science is necessarily neutral; so science is a bad thing. Qualms about the so-called neutrality of science jeopardize the supporting public attitudes without which science cannot thrive. Nevertheless, the logical version of the neutrality thesis deserves to be taken seriously, if only for its practical consequences. If neutrality in its logical sense has to be accepted, if science really has no intrinsic tendency to further human welfare rather than to increase human misery, the case for massive and indiscriminate support of science may well appear problematic. Even if the human usefulness of science can be grounded in its *indirect* bearing upon welfare and happiness, it is natural for a firm believer in the neutrality thesis to have serious qualms about the value of science. A layman might well ask: If scientific truths are only bridges between non-rational directives and evaluations, what is so *good* about science? If a scientist is really, from a logical point of view, a kind of *engineer*, supplying factual bridges between the normative and evaluative considerations that matter most to human beings, what is so good about being a scientist? (The roots of one kind of anti-scientific revolt – which I deplore – may be found in this crude kind of attitude.) Questions about whether science as a whole, or whether particular scientific projects are worth pursuing are, themselves, not scientific questions, but they cannot be brushed aside. For science, *considered as an activity,* is not an uncontrollable natural phenomena like the earth's rotation. As a system of voluntary activities it is, at least in principle, amenable to internal or external control.

Neutrality and the subjectivity of norms and values

What I shall call "the neutrality thesis" becomes particularly pernicious when coupled with a correlative view about the alleged "subjectivity" or normative and evaluative judgements and especially so when the realm of rationality is exclusively identified with that of scientific discourse. Let me give two recent examples of this. In an article entitled "On the Logical Relationships Between Knowledge and Values,"[2] the well-known biologist, Jacques Monod, says the following:

> Science rests upon a strictly *objective* approach to the analysis and interpretation of the universe, including Man himself and

> human societies. Science ignores, and must ignore, value judgements. Knowledge (yet) discloses and inevitably suggests new possibilities of action. But to *decide* upon the course of action is to step out of the realm of objectivity into that of values which, by essence, are non-objective and therefore cannot be derived from objective knowledge. There is strictly no way of *objectively* proving that it is BAD to make war, or kill a man, or to rob him, or to sleep with one's own mother.

On the same page Monod says "Objective knowledge defines the alternatives. By itself, it does not in any way, help in solving the agonizing ethical problems which science only poses." Later on, he says "Science indeed cannot create, derive or propose values" and he speaks about the necessity for the *"choice"* of value systems. Even the decision to become a scientist, says Monod, results from a "deliberate axiomatic choice of a standard of values".

In these remarks, which could be paralleled by any number of similar ones by distinguished scientists and intellectuals, objectivity connotes what is "given", regardless of human wishes and valuations – while "values" are treated as a matter of mere arbitrary "choice." So the questionable but unquestioned dogma of the gulf between the factual and the normative gets coupled to an equally dubious dogma of the arbitrariness and subjectivity of all norms and values.

My second example is taken from an article published only a week ago in the *New York Times Magazine.*[3] Heilbroner, a professor of economics at the New School for Social Research, there considers the questions, once raised by Adam Smith, whether "a man of humanity, given the choice, would prefer the extinction of a hundred million Chinese in order to save his little finger?," (Why Chinese, one wonders? Would it have made any difference if the lives of a million Englishmen had been in question?) One might have supposed the answer to Adam Smith's question was obvious. But listen to Professor Heilbroner's comment. He argues that it would be "outrageous" to attach such fantastic value to one's finger (or, to turn to his more immediate subject, outrageous to sacrifice posterity for the benefit of those alive today). But he thinks that "there is no rational answer "to such terrible questions. And this according to Heilbroner "reveals the limitations – worse, the suicidal dangers – of what we call 'rational argument' when we confront questions that can only be decided by *an appeal to an entirely different faculty from that of cool reason.* More than that," he continues "I suspect that if there is cause to fear for man's survival it is because the calculus of logic and reason will be applied to problems where they have as little validity, even as little bearing, as the calculus of feeling or sentiment applied to the solution of a problem in Euclidean geometry." Here we have a kind of muddled fantasia on a theme from Hume. "cool reason" is contrasted with "the unbearable anguish we feel if we imagine ourselves as the executioners of mankind." Well, if Heilbroner is prepared to rely upon the "unbearable anguish" of responsibility for mortal injury to human beings, he cannot be an attentive reader of the distinguished newspaper in which his article appeared. Can anybody seriously think there is *no good reason* for preferring the loss of one's little finger to the murder of even a *single* human being? Or that people should be regarded as perfectly rational if they were to adhere to an insane scale of values? I can only marvel at the advocacy of so peculiar a notion of rationality. If prevalent conceptions of rationality confine reason to the consideration of the factual and technical propositions, so much the worse for such restrictive conceptions of rationality. Scientific activity regarded as rational in *this* way, runs the danger of speedily becoming *inhuman.*

I have cited these examples, not to pillory the eminent gentlemen in question, but merely to illustrate the pernicious effects of the dogma of the absolute separation between 'is' and 'ought,' between the factual and the normative.

Science as activity

So far I have been discussing science in its relations to values and norms in a very abstract way, as if scientific propositions were simply given like natural objects, and as if the only proper questions for consideration were the logical ones of the relations between scientific propositions and normative or evaluative ones. It is time to remind ourselves of how much is ignored in this abstract approach and to consider science as a *system of human activities* – a far more than merely contingent or incidental contrivance for the delivery of "eternal truth." So we must raise the question of "neutrality" again in connection now with the enormously complex system of activities that deliver what for the time being at least counts as "knowledge."

What are we to say about the neutrality of science considered as a complex of *actions*? Well, the most obvious point is that *every* human action is in principle subject to practical and moral evaluation. It always makes sense to ask whether what human beings are *doing* is prudent or wise, satisfactory, intelligent, right or wrong. All action is directed rightly or wrongly, correctly or mistakenly, towards ends held by the agents to be right. Accordingly, it always makes sense to ask whether these ends are properly evaluated or whether there is reason to suppose that their choice has been misguided or evil. Scientists, like all other human beings, in their capacity as active *agents* are necessarily subject to such normative and evaluative appraisals.

We are therefore entitled, indeed obliged, to ask such questions as: What are the *goals* of science? Can they be approved without qualification or reservation? (Are they supreme?) What are the costs ("opportunity costs" as economists say) of supporting science, or some particular scientific project, rather than doing something else? The only ways in which these questions could be brushed aside would be if the supreme goals of science were either plainly and uncontroversially good (or at least universally accepted as such) or on the other hand if the supreme goals of science were so unimportant, humanly speaking, as not to deserve scrutiny. Similar considerations apply to the *means* used by scientists: issues of moral evaluation could be brushed aside if the consequences of scientific activity were so trivial as to be negligible and not worth discussing, or so obviously good as to make criticism pointless. In short, the only way to dodge moral questions about the values of science would be to hold that the goals and procedures of science were either practically speaking inconsequential or else uncontroversially good.

My original question about the neutrality of science, now that we focus upon the manifestation of science as human activity, becomes first, "Is science negligible or harmless in its consequences?" and secondly, "Are the ends of scientific activity unequivocally and unreservedly good?" We are led therefore inescapably to consider the practical importance of science and what might be called its evaluative dignity.

I shall not linger over questions concerning the practical momentousness of scientific activity, since that is too obvious to need discussion. But it is worth raising the question about the final end of scientific knowledge because, as I shall agree, there is a tendency to transfer illicitly certain important and genuine values, implicit in the disinterested pursuit of knowledge for its own sake, to the *whole* of scientific activity, in all its amplitude and variety.

Science as the disinterested pursuit of knowledge

Science, as its etymology and that of its German equivalent, "Wissenschaft", indicate, does have some essential connection with *knowledge.* Considered as an activity, it needs to be considered rather as the *pursuit* and the *discovery* of knowledge and has of course been so defined on innumerable occasions. It is perhaps a pity that instead of the fairly new word "scientist" we do not speak, in the old style, of scientific investigators as *natural philosophers,* human beings engaged in the *pursuit* of knowledge.

If we are to consider whether the pursuit and love of knowledge is a supreme and unqualified good we shall have to make an obvious distinction between knowledge that issues in the form of *information* and knowledge that yields *understanding* – between the collection of items of fact and the search for explanatory principles (at all levels of depth from bare empirical generalization to the deepest principles of physics).

It seems obvious to me, and I hope will be so to you, that knowledge in the form of information is *not* without qualification a good thing.

There is probably more truth (more "bits" of information) in a new telephone directory than in the whole of von Frisch's beautiful work on bees; but one would have to have a perverted taste to prefer the former to the latter or to advocate "scientific" research on the contents of telephone directories. If the gathering of information were an unqualified good we would have to salute the Watergate conspirators and the keepers of the CIA files as behaving like true scientists in their somewhat bumbling fashion.

Apart from such obvious points, there is the consideration that the effective and successful handling of knowledge demands the constant suppression and elimination of *irrelevant* information. To see anything at all is necessarily to turn a blind eye to what you don't want to see: if you look at individual trees you will be unable to see the wood; and if you look at the wood you will necessarily ignore details about the trees. The collection and systematic accumulation of knowledge is inescapably *selective.* Because it requires involving human *choices* it therefore inescapably commits the chooser to making value judgements. Therefore, even in its most prosaic function of fact-collecting, scientific activity is bound up with judgements of value. There is no particular virtue in collecting trivial information: the pebbles on a million beaches are waiting to be counted but one would have to be a fool or an advanced thinker to identify that kind of pointless fact-gathering with the exercise of admirable scientific activity.

When we turn to the other supreme product of scientific activity, *understanding,* we might suppose the implicated values are obvious. Isn't it always *better* to understand than to be blinded by illusion or prejudice? Well, the same elementary points apply: it all depends upon the *objects of understanding,* it all depends upon their value and the uses, if any, which the understanding is to serve. Is it to be seriously argued that it is a good thing to understand the most painful ways of killing and torturing innocent people – just because it is *knowledge*? Would it not be better to *be* and to *remain* ignorant about such matters?

The chief point that I am making, and I hope a fairly obvious one, is that the worth and undeniable dignity of science, considered as the disinterested pursuit of knowledge, whether of facts or of explanatory principles, depends upon value choices – choices of significance and relevance, choices of importance and fruitfulness, and above all, choices concerned with *ultimate worth to human beings.* Paul Goodman once said, in a striking phrase, that "technology is a branch of moral philosophy, not of science."[4] I am not sure that I agree: moral philosophy is difficult enough without trespassing upon engineering. But one can see what he means. I suggest that the same insight also applies to *science,* even in its purest manifestations. Considered in the round, the "purest," the most abstract and apparently "useless" branches of science are inescapably value-dependent and value-laden.

I do not wish to be misunderstood as an enemy of basic science or an enemy of the pursuit of science for its own sake. There are plainly great values in disinterested science, properly enlisting the devotion and sacrifice of some of the most gifted members of the species to which we belong. The achievements of Galileo and Newton, of Darwin and Einstein, to name no others, add to the stature of all of us. And the disinterested pursuit of truth about the physical universe, the biological realm and the political and social institutions is a great value, hardly won, constantly under attack, and to be secured only by constant vigilance. I share the view of Julien Benda in his famous book, *Le trahison des clercs* (translated as *The Great Betrayal*) that we shall always urgently need truth seekers, not only scientists, but scholars and artists and poets as well, who will pursue and declare the truth as they see it, without regard for factional and national interests or for the distorting influences of passion, envy and greed.

But the autonomy and freedom from all interference or control that is part of the ethos of the disinterested truth-seekers can be granted only upon the condition – that he or she is genuinely interested in knowledge and understanding for its own sake and nothing else. In short, that contemplative knowledge of the universe be indeed a common good accessible to all, shared and enjoyed by all and not something that is deleterious to human welfare. Now the great dilemma posed by the extraordinary and unforeseeable efflorescence of science and technology has been the increasing impossibility of separating the noble and sublime conception of science from its practical consequences. Science has become a terrible two-edged sword, a power for great evil as well as for great good, and this puts the question about the 'neutrality" of science into an altogether new perspective.

Science today

What might be called the *classical ideal* of the scientist as a disinterested pursuer and lover of truth can be wholeheartedly followed by *solitary* scientists or *solitary* scholars, Darwin in his study or Gilbert White in the countryside around Selborne, or even Einstein in the Patent Office at Berne, calmly discovering and contemplating truth, with *no ulterior motive,* for the sake of the radiant pleasure of the pursuit itself, isolated from the marketplace and the war-room, sharing their discoveries only with a small group of like-minded devotees of the truth for its own sake. But this idyllic picture, which can hardly have been a reality except occasionally, has long been overtaken by the Baconian ideal of science for the sake of "dominion over nature" – for the sake of "the effecting of all things possible."

We have come a long way from the classical ideal of the independent scholar communicating only with a small group of like-minded researchers. The latest official figures that I have been able to obtain show that in the United States alone there were in 1970 well over a quarter of a million persons identified as practising natural scientists and nearly 60,000 social scientists – altogether some 313,000 to the nearest thousand. If we count engineers and natural scientists, working in research and development, the total rises to something like 1,600,000. For the cost of this extraordinary effort we have an estimate for 1973 of over 30 billion dollars for research and development of which 11 billion dollars were for basic and applied research. The Department of Defence itself spent in that year over 8 billion, for research and development.[5]

Considering now the intellectual product alone, we find such statistics as these: that the *Physical Review* for 1973 amounted to 30,600 pages and weighed approximately 146 lbs. For the same year, the number of abstracts published in *Physics Abstracts* amounted to 81,352 and occupied 5,302 pages.[6] My estimate is that anybody who undertook simply to read superficially the volume of work produced in physics in that one year alone would have to spend anything from 3 to 6 months doing nothing else at all.

Now it is implausible that society should spend such enormous sums, and promote such intensive training and employment of high talent, in the service of the noble ideal of the disinterested pursuit of truth. Indeed, considering the very high proportion of financial support for science that comes from industry and departments of defence, one would expect the *effective* goal of this vast enterprise to be at least strongly compatible with the ruling goals of industry and the military, that is to say, that of providing ways of making more profit and ways of making more efficient instruments of destruction. This intimate coupling of scientific activity with the industrial-military complex has cast a shadow over the noblest aspect of Bacon's vision, the use of science to increase human welfare and human happiness. Fortunately, welfare and happiness can often be profitable, so a more sanguine view of the situation might dwell upon the undoubted and enormous contributions of applied science and technology to the control of disease, improvement in communication, and the pleasure and comfort of say 20 per cent of all human beings alive today.

The responsibility of science

But the plain lesson of the fantastic explosion of science in modern times is that the separation of pure science from its applications is no longer feasible in the context of moral evaluation. The purest, the most recondite discoveries of disinterested scientists can now, in a relatively short time, be harnessed to the production of new materials, new contrivances, new ways of transforming individual lives and the very shape and structure of society.

This means that the ground has been cut away from the classical claim for the autonomy of pure science. So long as the scientist could be regarded as a harmless discoverer and contemplator of aesthetically satisfying truth, the rest of us could admire and applaud, even where we could not understand, and consider humanity dignified by such devotion, as it is dignified by the existence of any spectacular realization of human excellence. But the case is altered when the consequences of such admirable activity can be and are often intended to be of immediate and momentous effect, so momentous that the very existence of what is essential for the bare survival of human beings becomes a debatable and unsettled question. Science is even more terrible in its potentiality for evil than the atomic bombs for which it is responsible. Over every scientific laboratory should be pasted the warning: "Danger: Scientists at Work".

Whether one takes an optimistic or pessimistic view of the probable consequences of the massive transformation of the natural and man-made environment and the conditions of social life that are traceable in the end to the innocent classical delight in knowledge, the conclusion is inescapable, that the claim of moral innocence must now be dismissed as a mere vestigial survival of an earlier and safer era. It is one thing for a pure geometer to demand to be left alone, without impertinent claims from the surrounding society to prove that his work has any practical application, and without irrelevant injunctions to pay attention to the negligible immediate consequences of what he is doing. (To borrow Samuel Johnson's phrase, one might at worst call him a "harmless drudge".) It is quite another thing to hear the same plea of moral blamelessness and "neutrality," in the sense now of moral *innocuousness,* made by contemporary scientists working directly, say, for the military establishment. When the immediate consequences of scientific activity are so plain and so obvious, a plea of "scientific neutrality" can only be properly characterized as an expression of deliberate myopia or, to put it bluntly, *moral irresponsibility.*

An example of moral irresponsibility

Allow me to offer an example of this kind of irresponsibility. In *The New York Times* for December 27, 1967 (page 8), there appeared an interview with Dr. Louis Frederick Fieser who was in charge of a team of Harvard University scientists developing napalm during World War II. He is reported as having said that he he felt free of "any guilt." He is also quoted as saying "You don't know what's coming. That wasn't my business, *That is for other people.* (my italics) I was working on a technical problem that was considered pressing." He went on to say: "I distinguish between developing a munition of some kind and using it. You can't blame the outfit that put out the rifle that killed the President. I'd do it again, if called upon, in defense of the country." When asked about the use of napalm in Vietnam, Dr. Fieser said, "I don't know enough about the situation in Vietnam. It's not my business to deal with political or moral questions. That is a very involved thing. Just because I played a role in the technological development of napalm doesn't mean I'm any more qualified to comment on the moral aspects of it."

The moral callousness of these remarks is matched by their confusion of thought. The idea, for example that one needs moral "qualifications" in order to consider the consequences of one's action is very odd. But it is unfair to saddle Dr. Fieser alone with this attitude, which continues to be widely held. The social scientist, George A. Lundberg, said it plainly, nearly 40 years earlier.

> "It is not the business of a chemist who invents a high explosive to be influenced in his task by considerations as to whether his product will be used to blow up cathedrals or to build tunnels through the mountains. Nor is it the business of the social scientist in arriving at laws of group behavior to permit himself to be influenced by considerations of how his conclusions will coincide with existing notions, or what the effect of his findings on the social order will be." *Trends in American Sociology.* Edited by G. A. Lundberg, R. Bain and N. Anderson. New York, Harper, 1929, pp. 404–405.
> (quoted from Merton, Robert K. (1953). *Social Theory and Social Structure,* Glencoe, Illinois. The Free Press. p. 543.

Against much misplaced high-mindedness, we ought to say that in the case of research plainly intended to produce flamethrowers and incendiary bombs the moral responsibility of the chief inventors and artisans cannot be evaded by professions of innocence. (That the persons most directly involved can have a good conscience only adds to the irony – might one even say the horror – of the situation.) Responsibility for the invention and use of such "scientific devices" as napalm is of course shared by many – and no doubt belongs primarily to those who have the final power to commission and to use such inhuman devices. But the inhumanity of statesmen does not absolve technologists, scientists and all of us from complicity. And given the key role played by scientists at all levels in involvement it is particularly hard to absolve them from fairly direct responsibility. As I said at the outset of this talk, my purpose has not been to preach. Nor has my purpose on this occasion been to suggest practical measures by which to discharge the undoubted responsibility that is borne by science and every society that supports it. My purpose has been the logically prior one of arguing that the responsibility exists and will be discharged one way or the other, whether by the determined effort of scientists and laymen who are not content to transfer the ideas of an earlier age to the present situation or to regard scientists as mere hired help fit for any purpose, good or evil – or else by others somewhat less hampered by any appreciation of the value of disinterested research. The optimist can hope for a time when science will have returned to its ancient and nobler functions of enlarging human understanding and relieving human misery. But as Groucho Marx must surely have said upon some occasion, "It takes an optimist to be an optimist nowadays."

You may reasonably expect me, in conclusion, to answer the question that I posed an hour ago. Is scientific neutrality a myth? I must answer, I am afraid, in the irritating style of a philosopher: "It all depends upon how you conceive of science." (And, of course, upon how you conceive of neutrality – but I shall say no more about that.)

If you are satisfied to abstract from the extraordinarily ramified complex of actions, institutions, traditions, together with the supporting educational, economic, industrial, and political structures that we comprehensively denote as "Science" that aspect of it that consists of the "disinterested pursuit of knowledge" for its own sake, scientific neutrality is *not* a myth, but a justified *claim.* Nothing that I have said should be misconstrued as a plea for political or any other interference with the intrinsic demands of scientific investigation.

But if you think such emphasis upon the disinterested investigator is too partial, too abstract a view of the whole of science to be serviceable, if you think of science in its totality of present-day activities, I think you will need to reject the neutrality thesis as a myth and a pernicious one. For a particularly dangerous kind of myth is one which, under the guise of truth, insinuates a claim to moral irresponsibility.

Footnotes

1. Hume, David. *A Treatise of Human Nature,* Book 3, Part 1, Section 1. Those who wish to pursue the philosophical issues further might be referred to my essay, "The Gap Between 'Is' and 'Should' ", Chapter 3 of *Margins of Precision.* Ithaca, N.Y. Cornell University Press, 1970.
   A useful collection of related papers is W. D. Hudson (ed.). *The Is/Ought Question.* London, Macmillan, 1969.
2. Included in Watson Fuller (ed.) (1971). *The Social Impact of Modern Biology.* London, Routledge & Kegan Paul.
   The quotations I have used will be found on pp. 12 and 15.
3. "What has Posterity Ever Done for Me?", *New York Times Magazines,* 19 Jan. 1975, pp. 14–15.
4. In his essay, "Can Technology Be Humane?", quoted from Martin Brown (ed.) (1971). *The Social Responsibility of the Scientist.* New York, The Free Press, p. 251.
   The quotation continues "It (technology) aims at prudent goods for the commonweal and to provide efficient means for these goods."
5. Figures taken from the *Statistical Abstract of the United States, 1974,* prepared by the U.S. Department of Commerce.
6. These data were obtained by my colleague, Bob Linden.

Appendix Two
Guilty until Proven Innocent

By Garret Hardin

It is arguable whether being a king in the old days was preferable to being a commoner, most of the time; but when it came to dying there is no doubt that a king had the worst of it. The trouble was that he was given the best medical care available in his time. His commoners were lucky enough not to be able to afford it.

Consider what Charles II was subjected to, as he lay dying in 1685.

> A pint of blood was extracted from his right arm; then eight ounces from his left shoulder; next an emetic, two physics, and an enema consisting of 15 substances. Then his head was shaved and a blister raised on the scalp. To purge the brain a sneezing powder was given, then cowslip powder to strengthen it. Meanwhile more emetics, soothing drinks and more bleeding; also a plaster of pitch and pigeon dung applied to the royal feet. Not to leave anything undone, the following substances were taken internally: melon seeds, manna, slippery elm, black cherry water, extract of lily of the valley, peony, lavender, pearls dissolved in vinegar, gentian root, nutmeg, and finally 40 drops of extract of human skull. As a last resort bezoar stone was employed. But the royal patient died.

Died of what? Of the medical treatment itself more than likely – an "iatrogenic" death, as the medical profession delicately puts it ("physician-generated", literally). Kings could afford such a death; commoners could not. The biochemist L. J. Henderson reckoned that until about the year 1905 calling in a physician to attend the sick *decreased* the patient's chances for survival; after that time medical attention gradually became of positive value. With the coming of sulfa drugs in the 1930s and the antibiotics in the 1940s, the balance swung definitely in favor of medical treatment.

We deride the flounderings of the physicians treating King Charles's ailment (whatever it was): but are we any more intelligent or effective in treating the ills of society? Posterity may conclude not. In both instances an ailing and little-understood system is treated as if it were merely a state that was awry. The measures used are all too often merely a form of sympathetic magic: removing blood to cure high blood pressure, or applying money to cure poverty. As Oscar Wilde said of the social do-gooders, it is much easier "to have sympathy with suffering than it is to have sympathy with thought. Accordingly, with admirable though misdirected intentions, they very seriously and very sentimentally set themselves to the task of remedying the evils that they see. But their remedies do not cure the disease: they merely prolong it. Indeed, their remedies are part of the disease."

That was written in 1891. Are we more successful today? Have we, in treating the ills of the body politic, yet passed the equivalent of Henderson's 1905 line for human medicine? Do we yet have the capability of doing more good than harm by social interventions that *intend* good? ("But we just have to do *something*!" says the do-gooder when faced with suffering. But suppose that that something increases suffering? Are good intentions then an acceptable excuse?)

When ignorant, what is the best thing to do? If a would-be physician offers us a nostrum, whether for the human body or for the body politic, should we try it? Should we be adventurous or conservative? Should the peddler be allowed (or encouraged) to peddle his pills?

Where does the burden of proof lie? This is the fundamental question. In criminal law, as practised in Britain and America, a man is "innocent until proven guilty." Quite naturally this policy has been carried over to the realm of medical nostrums. It is in the interest of the peddler that this extension of the law is made. Patients, clutching desperately at straws, often acquiesce (and if they die, their second thoughts don't matter).

Scientists, however, see things otherwise. Science is an occupation in which most experiments fail. Those who cleave to science in the face of constant disappointment expect failure and disappointment as a matter of course (though naturally they hope that the usual outcome will fail to appear once in a while). Confronted with any new, untried nostrum, a scientist, if called upon to place a bet, will bet that it won't work. Such is the conservative judgement.

The overwhelming probability is that any newly proposed remedy won't work. More: experience shows that there is an almost equally high probability that the new nostrum will cause actual harm.

The most intelligent way of dealing with the unknown is in terms of probability. Therefore we should assume that each new remedy proposed will do positive harm, until the most exhaustive tests and carefully examined logic indicate otherwise.

Guilty until proven innocent – this should be our assumption regarding the value of each newly proposed remedy. The law regarding remedies should be sharply differentiated from the law governing human beings accused of crimes. Although understood for decades by almost all scientists and a great many laymen, the principle of "guilty until proven innocent" was not given effective legal status in the United States until 1962. The change would not have been made even then, in all probability, except for the fortunate tragedy – if such a paradoxical expression is permissible – of the thalidomide babies. The newly discovered tranquilizer thalidomide was used by millions of people before one of its dreadful "side-effects" was discovered: taken by a mother in early pregnancy, it produced badly malformed babies, typically with stumps for arms and legs. Thousands of deformed children were born in Europe before the causal connection was uncovered in America, where the drug had not yet been approved for release. The delay on this side of the water was probably due less to prescient suspicions than it was to normal bureaucratic delays. There's something to be said for inertia and conservatism.

In the wake of the thalidomide tragedy the Kefauver-Harris amendments to the Food, Drug and Cosmetics Act were passed in 1962. These amendments put the burden of proof on the proposer of a new remedy, who, before his product could be licensed for distribution, had to show that it was:

a. effective
b. harmless (or, more exactly: did more good than harm).

This was a revolutionary change in the assumptions of the law. That it should be made first in the area of medicine is understandable: the common man's fear of sickness and death is so intense that the usually effective cries of "socialistic," "un-American" and "a step down the path to the police state" fall on deaf ears. "Better dead than Red"? Not in the mind of John Q. Citizen when he is confronted with the real possibility of death. Moral perfection in death is a luxury most men can do without. In 1962, through their elected representatives, the American people decreed that there are some areas in which they are willing to assume "guilty until proven innocent." Will this attitude spread to other fields? We shall see.

Bringing about an administrative revolution is more difficult than merely passing a revolutionary law. Only gradually did the Food and Drug Administration alter its attitude toward newly proposed nostrums. Administrative minds had been too long biased in favor of the profit motive. Not until about 1970 did retirement and reform bring the FDA personnel to the point of carrying out the aims of the Kefauver-Harris amendments in the vast area of re-evaluation of old remedies.

A European pharmacopoeia of the seventeenth century listed some 6000 drugs in use. In 1960 a similar list in the United States boasted about 3000 drugs, in over 10,000 different formulations. Not much of a reduction in three centuries. Undoubtedly most of the 3000 nostrums are worthless; probably most are more or less harmful. But they were protected by tradition until the Kefauver-Harris amendments were directed at them. Now they must stand on their own feet – i.e. pass a gauntlet of rigorous and objectively controlled tests before they can continue on the approved list.

The FDA has begun to remove whole blocks of traditional remedies from the approved list, to the acute discomfiture of drug houses, which are fighting back. Some of the delaying tactics are successful. The long-term results will be determined by the extent to which the scientific attitude has permeated the public mind. Those who understand the connection of probability with intelligent action in the face of the unknown are necessarily conservative and assume *guilty until proven innocent.*

A word of warning is in order. It is always dangerous to take a rule that is good in a limited context and generalize it widely without careful point-by-point testing. The doctrine of "innocent until proven guilty" is no doubt the best of all general rules in criminal law, where the treatment of *people* is at stake. Extending it to *things* led to widespread deception of customers. It would be at least as unwise to extend the doctrine of "guilty until proven innocent" from the area of things to that of people (i.e. to the criminal law). Generalizations should not be escalated.

The passage of the National Environmental Policy Act of 1969 (enacted into law on January 1, 1970) moved the nation significantly closer to applying the guilty-until-proven-innocent principle to the environmental realm. It did not go so far as to spell it out in general, but it did specify that any proposal of any agency of the federal government must be examined in minute detail for its effects on the environment.

Later in 1970 a small step was made toward applying the same principle to nongovernmental agencies. Henry Reuss, a Congressman from Wisconsin, had unearthed a still valid law governing the discharge of materials into inland water, the 1899 Refuse Act. Surprisingly, considering its antiquity, this law said that nothing can be discharged into a navigable water without federal permission. Needless to say, consternation greeted the discovery of this old act. To enforce it immediately would be impossible both practically and politically. A compromise scheme was put forward.

The Army Corps of Engineers was empowered to issue effluent licenses. Were it not for the changing temper of the times, this arrangement would have been rather like setting the fox to guard chickens; but the corps is changing and it worked out an agreement with the newly established Environmental Protection Agency whereby the EPA is to advise the corps on major requests. Because of the sheer volume of the new monitoring to be done, drastic changes in effluent discharge should not be expected soon; but the requirement that businesses reveal what they are discharging, and how much, moves industry a giant step toward initiating rigorous accounting for the flow of materials through its plants. A foundation is laid for environmental decisions to be made later from the standpoint of the general public.

Like the human body, the environment is an enormously complex system of interacting elements and processes – and most of the interactions are unknown. Disrupt this web of life with a random intervention: What is the probability that harm will *not* be done? It is surely vanishingly small. The conservative approach is, therefore, to make no change at all without an exhaustive investigation first.

That is not the way we have treated the environment in the past. Take as an example the detergent problem. We used to wash clothes with soap. It got them clean enough, and caused no great harm to the environment. But we didn't leave well enough alone: we replaced soap with detergents. In terms of the single measure "getting clothes clean," this was an advance. Detergents got them cleaner, and they worked better in hard water than did soap.

Unfortunately the detergents lasted almost indefinitely. Discharged into the streams and the ground water, they proved unsuitable as bacterial food and so they accumulated. Ponds below dams were often covered by billowing mountains of froth. It wasn't certain that detergents were particularly dangerous, but when a glass of water drawn from the tap had a "head" on it like a stein of beer, the public became alarmed. People did not like to be reminded that they drink sewage, more or less modified by time and treatment, but sewage nonetheless. The "hard detergents" – called "hard" because they are hard for the bacteria to decompose – made it difficult to forget the origin of the drinking water.

Faced with a mounting public outcry, the cleanser companies, at considerable expense, changed from hard detergents to soft. The soft detergents had phosphate compounds in them. Phosphates are required by all living things. Soft detergents are easily decomposed by bacteria, releasing phosphate ions that are utilizable by anything that lives. They're food . . . It looked as though the cleanser industry had found the perfect solution.

Unfortunately, it's possible to have too much of a good thing. Our cleanly housewives require about two and a half million tons of detergents a year which (on decomposing) add about a million tons of phosphates to the ground water, the streams, the lakes, and coastal estuaries. That's a lot of phosphate. Phosphate is quite often the limiting factor in algal growth, so if you double the phosphate concentration you double the amount of algae. Ten times as much phosphate ten times as many algae.

Is that bad? Not necessarily. Algae are part of the food chain, so a small increase in phosphate can increase the production of desirable fish. But when the increase is large, the picture changes. Great mats of algae are produced, and the species composition of the algal mats swings over in favor of blue-green algae, which are not readily eaten. The mats grow so thick that they cut out the sunlight to the lower algae, which die, setting up anaerobic conditions that favor more death. The water stinks and fish die. "Eutrophication" sets in – "perfect feeding," literally, but the practical meaning is perfectly terrible over-feeding of the aqueous world.

As they perceived the disadvantages of phosphates the cleanser-men looked around for something else. This time they came up with enzymes – digestive enzymes gotten out of fractured bacterial cells, which can easily be grown by the millions of millions. The proteolytic enzymes – protein-digesting – were touted for their ability to remove resistant stains, of blood for example. Enzymes were added to other cleaning agents. Whether they were or were not effective is disputed, but the dispute soon became of only academic interest because it was discovered that the new formulation had a very serious defect.

All enzymes are proteins. Foreign proteins – proteins derived from flesh other than human – can produce allergic reactions in humans who are repeatedly exposed to them. There are great individual differences in susceptibility to allergic reactions, but if millions of people are exposed to an allergen, thousands will react adversely. Soon after the introduction of enzyme-detergents, housewives started turning up with rashes and respiratory symptoms. Workers in the factories where the detergents were packaged were even more affected, being more exposed.

What next? Next came NTA, or nitrilotriacetate, to give it its full name. No phosphate in it. Unfortunately there is some evidence – not conclusive – that it causes birth defects, perhaps only when combined with heavy metal pollutants like cadmium or mercury, which (however) are not unknown in our environment . . . Obviously the cleanser-men never thought of *that.*

What's the moral of this long and unfinished story? NTA has been phased out (before being really phased in), and phosphates have been retained while the industry looks for another chemical compound.

Why don't the cleanser-men test the compounds before they market them? *Test them for what*? There's the rub. They have R & D departments – Research and Development – that are constantly testing, but the focus of their investigations is narrow, Newtonian: to find something that will clean. They are not public-health researchers; they are not ecologists: they are not systems analysts. It's never occurred to them that a good cleanser might, in time, raise rashes; or cause birth defects; or kill fish; or put a head on tap water. These are not obvious eventualities: why should anyone even think of looking for them?

The next chemical candidate for the cleanser market will no doubt be checked for all these evils – but it may well turn out to have yet another disadvantage that no one dreamed of looking for.

The cleanser industry R & D is looking for just one thing: a better way to clean. But we can never do merely one thing – this is the basic ecological wisdom. Every innovation is an intervention in the entire ecological system. Ideally, we should check every one of the millions of elements in the web of life, every one of the unnumbered trillions of relationships, before adopting any innovation. This is impossible. So we check a few of the most obvious and then hold our breath as we make the change. A small pilot-plant experiment may fail to reveal results that appear only when the change is expanded to a large scale. Perhaps only a few per cent of the people exposed may be affected – but a few per cent of a population of 205,000,000 is millions. Or the adverse affects may take time to develop – and time inevitably passes.

Just thinking about it is enough to make a technology-liberal into a technology-conservative. At times one wonders if one shouldn't go back to using soap. Or pounding clothes on a rock in the creek.

In this, as in almost every other unsolved problem of our time, population plays a role. Suppose the United States had only twenty million inhabitants instead of over two hundred million. Remembering that phosphates are a good thing if not too concentrated, we realize that with the lower population density there would probably be no reason to become concerned with the disadvantages of new detergents. We would know only the advantages.

The disadvantages of each new intervention in the web of life increase with population size. Recycling can remove the pollutant – but not if its concentration passes a critical value, as it will at some level of population.

A small population can be cavalier in its assessment of the disadvantage of new technology, and get away with it. Not so a large population.

The larger the population grows, the more conservative it necessarily becomes in its attitude toward technological innovation. Either that, or it suffers more.

The developing shift in the assumption of the law has not slipped in without notice by commercial interests. Shortly after he came into office President Nixon established a National Industrial Pollution Control Council, made up of some threescore presidents and board members of the wealthiest corporations in America (hardly the most promising group to monitor life on a spaceship). In 1971 this group alerted the President to a dreadful danger they saw looming on the horizon: "The view that no material should be permitted to be released into the environment unless it can be shown to be harmless to man and his environment . . . could be likened to a requirement of proof of innocence by the accused rather than proof of guilt by accuser." Their (external) vision is excellent; what they signally lack is insight into the problem. What they abhor deserves praise. Will they ever see this? Considering that their average age is probably greater than sixty, and taking account of the fact that they are *very* busy men, one would guess not.